Unintended Consequences:
The Lie that killed millions and accelerated Climate Change
by George Erickson

Without the input provided by many members of the Thorium Energy Alliance, especially Dr. Alex Cannara, this book would not have been possible.

Unless otherwise credited, the images and charts in this book are from Imgur, Wikipedia or the Thorium Energy Alliance

Fair use statement

This book might contain material that has not been authorized by the copyright owner. I make this material available in an effort to advance understanding of environmental, economic, political and scientific issues.

Cover art by Spencer Hahne

Other books by George Erickson

True North: Exploring the Great Wilderness by Bush Plane

Time Traveling with Science and the Saints

Back to the Barrens: On the Wing with da Vinci & Friends

Eyes Wide Open: Living, Laughing, Loving and Learning in a Religion-troubled World

Contents

Foreword
Why I Care

Preface
A Deadly Evacuation

Chapter 1
Climate Change

Chapter 2
The Lie

Chapter 3
A Little Nuclear History

Beer and Bananas:

When Radiation is Safe and When It Isn't.

Chapter 4
DNA and Hormesis

When Radiation Can Even Be Good for You

Chapter 5

The Consequences of Overreaction

Alarming ALARA

Chapter 6

What's so Great about Nuclear Power?

Three Mile Island, Chernobyl and Fukushima

Chapter 7

The Fossil Fuel Record

Safety and Death-prints

Chapter 8

Powering Ships and Desalination

What's a Light Water Reactor?

Waste Management

What's a MSR? What's a LFTR?

Chapter 9

Blowin' Wind

Chapter 10

Solar and Biomass

Chapter 11

The Opposition

Anti-nuclear Zealots and Profiteers

Chapter 12

Recommended Reading

An Appeal to Reason

To receive a "clickable" list of the links that appear in this this book, email your request to tundracub@excede.net. Secondary address = caerick@mchsi.com.

Foreword

This book is intended to help open-minded readers learn the truth about nuclear power and alternative energy sources like wind and solar, and to explain why our fear of even tiny amounts of radiation has caused millions of deaths and disabilities.

However, those who challenge the firmly held beliefs of many legislative bodies and powerful organizations like the Sierra Club, the Nature Conservancy and their well-meaning, but science-indifferent clones, soon learn that their arguments, no matter how logical or well documented, will often be brushed aside with a dismissive "That's just your opinion."

To counter that assertion, I have included many links to supportive material from a wide range of professionals in the energy field: engineers, nuclear physicists, science journalists and specialists in nuclear medicine. Although inserting links to the work of so many professionals *within my text,* instead of footnoting them, might seem intrusive, I've taken that risk because the health of our planet *requires* an informed public and educated legislators, especially when the technologies they have chosen are damaging the environment that they claim to revere.

Unfortunately, as I tour the country with my climate change presentations that support advanced nuclear power, and criticize an efficient, anti-environment wind and solar, I am constantly reminded of Mark Twain's perceptive comment, "It is easier to fool someone than it is to convince them that they have been fooled."

This book is dedicated, with sincere apologies, to the
coming generations that will have to clean up
the mess that we have created.

Why I Care

Back in the sixties, when I was living in a small Minnesota farming community, my sons were taught to "duck and cover" beneath their desks in case of a nuclear war.

We'd also been warned about radiation and fallout, so I built a concrete block shelter in my basement that I hoped would shield my family for a week or two if events with Russia turned sour.

Time passed. The Cold War waned, and when concerns about nuclear power changed from making bombs to making electricity, thoughts of things nuclear receded - until I attended a lecture on thorium near the turn of the century. Intrigued, I began to investigate thorium because of its many advantages over uranium for producing electricity.

I joined the National Center for Science Education and the Thorium Energy Alliance, which provided a huge upgrade to my better than average knowledge of physics. And then came Climate Change.

I had known about greenhouse gases, global warming and sea level rise and I had read about Dr. Charles Keeling's work with carbon dioxide on the slopes of Mauna Loa, but I hadn't realized that expanding nuclear power, which creates no carbon dioxide (CO_2) could be our most effective weapon for combating Climate Change, of which a huge portion is produced by burning coal, oil and natural gas to supply electricity to an expanding world that exceeds 7 billion - a world that finally beginning to consider the value of CO_2-free, environmentally benign nuclear power.

One solution seemed obvious - keep the generators of the coal and gas-burning power plants, but replace their burners with nuclear power. However, I quickly discovered that powerful organizations oppose almost everything nuclear - some out of ignorance, many out of fear, and some for profit, but I also found support from those who'd set their fears aside after discovering the impressive safety record and efficiency of carbon-dioxide-free nuclear power.

And so, with Climate Change becoming deadlier every year (assisted by Donald J. Trump, our Climate Denier in Chief), and because my grandchildren's futures are at stake, I have decided to respond to those who fear our safest, most efficient, environmentally benign power technology by revealing its true record – including that of Chernobyl, where fewer than 70 have died, and of Fukushima Daiichi, where two people <u>drowned</u> at the plant - and I'll highlight some of the new nuclear plants that are even safer and more efficient than the hundreds we have relied on for more than 50 years.

But first, I must mention two discoveries that came as a huge surprise – the *fact* that our radiation safety standards are based on a fraud that became dogma not long after WW II, and the existence of compelling evidence that <u>low</u> levels of <u>background</u> radiation can even improve our lives. I know that sounds crazy. At first it did to me, but there is abundant science to back it up.

"An ecologist must be the doctor who sees the marks of death in a community that believes itself well and does not want to be told otherwise." Aldo Leopold – 1953

We must turn away from carbon.
We must do better than this!

Toles © 2013 The Washington Post. Reprinted with permission of UNIVERSAL UCLICK. All rights reserved.

Preface

A Deadly Evacuation

Excerpts from the Report of the United Nations Scientific Committee on the Effects of Atomic Radiation (UNSCEAR) 7-31 May, 2013 General Assembly Records

Chapter III Scientific findings [Fukushima]
"**1. The accident and the release of radioactive material into the environment.**

On 11 March 2011, at 14.46 [2:46 pm] local time, a 9.0-magnitude earthquake occurred near Honshu, Japan, creating a devastating tsunami that left a trail of death and destruction in its wake. The earthquake and subsequent tsunami, which flooded over 500 square kilometers of land, resulted in the loss of more than 20,000 lives. The loss of off-site and on-site electrical power and compromised safety systems at the Fukushima Daiichi nuclear power station led to severe core damage to three of the six nuclear reactors on the site…

The Government of Japan recommended the evacuation of about 78,000 people living within a 20-km (12 mile) radius of the power plant and the sheltering in their own homes of about 62,000 other people living between 20 and 30 km from the plant… *However, the evacuations themselves also had repercussions for the people involved, including a number of evacuation-related deaths and the subsequent impact on mental and social well-being.*"

Those "evacuation-related deaths" would eventually total 1600, with about 90% being caused by the Japan's reliance on American radiation safety standards that are based on a *fraud* that began in the 1930s. That fraud, committed by a Nobel laureate and formalized by the U. S. in the 50s, became regulatory dogma that would greatly retard the expansion of CO2-free nuclear power, accelerate Climate Change and cause the deaths of millions who, out of fear of radiation, avoided essential diagnostic methods and treatments that involved radiation, and at Fukushima, cause more than 1,000 suicides by distraught and unstable people, primarily the elderly, who feared that they would never see their homes or businesses again. (The daughter of an elderly woman who had hung herself lamented, "If she had not been forced to evacuate, she wouldn't have killed herself.")

Children were not allowed to play outside, and topsoil was needlessly removed at great expense from farm fields that became, as a consequence, less fertile. Hundreds of elderly people were hastily removed from nursing homes and hospitals, only to be scattered across the hardwood floors of gymnasiums, where many died from makeshift medical care, or sometimes none at all.

These deaths were preventable, just as climate change can be *moderated* if the industrialized nations *rapidly* replace the burning of carbon and the use of deadly, inefficient, carbon-reliant alternatives *like windmills and solar farms* (chapters 9 and 10) with CO2-free nuclear power as rapidly as possible while developing technologies that support natural processes that can remove CO2 from our atmosphere. Windmills can't do it. Neither can solar, not singly or combined with wind. For that, we will need an abundance of safe, efficient, CO2-free *nuclear* power.

Nothing else will do.

Chapter 1

Climate Change

In 1866, a Swedish chemist named Svante Arrhenius estimated that doubling our Earth's atmospheric CO_2 would raise its temperature by 9 degrees F, which is why CO_2 and its "associates" are called greenhouse gases (GHG).

Then, in 1958, Dr. Charles Keeling, an American chemist and oceanographer began to record the level of atmospheric CO_2 at Hawaii's Mauna Loa Observatory, which, being 10,300 feet above sea level and far out in the Pacific Ocean, avoided misleading data from mainland sources that would skew his research. Although Keeling eventually proved that CO_2 levels were soaring, few paid attention, and his work had little influence for more than 20 years.

The best part of the Mauna Loa road. (1985)
The remainder required four-wheel drive.

Acting like blankets, greenhouse gases limit how much of the Earth's heat can escape into space. If the blanket becomes too thin for too long, too much heat escapes, and an Ice Age follows. However, if it thickens excessively, too much heat is trapped, and the Earth develops a fever.

If we give water vapor a rating of 1, carbon dioxide would rate a 5, but methane, (CH_4 – the primary component of natural gas), is initially about 75 times more potent than CO_2 averaging some twenty times worse as it slowly oxidizes to CO_2 and H_2O. However, despite the fact that carbon dioxide is 5 times more potent on a molecule to molecule basis, water vapor is a more powerful accelerator of climate change than CO_2 because there is a lot more water vapor than CO_2, and as the planet warms, even more is created. The extra water vapor traps more heat, which raises ocean and land temperatures even higher.

For millions of years, our planet has been nurtured by a gassy comforter that, like Goldilocks' bed, has been just right. Those gases have served us well, especially since the last ice age, many thousands of years ago, varying only a little while periodically providing nothing worse than a string of challenging winters or abnormally hot summers before returning to normal. Fortunately, our blanket never varied enough to cause us significant trouble, but that has changed.

Thanks to air bubbles trapped in ice from Greenland and Antarctica, we know that the level of atmospheric CO_2 has been hovering near 280 parts per million (ppm) since the age of the dinosaurs. However, that number slowly began to rise about 200 years ago, when the Industrial Revolution allowed us to burn increasing amounts of carbon.

By 1950, atmospheric CO2 levels had reached 300 ppm. Spurred on by rapidly increasing industrialization and burgeoning populations, that figure reached 400 ppm in 2015. Now, though hampered by an anti-environment president, his carbon-loving, anti-science cabinet and a distracted, science-deficient Congress, we must put planet before profit if the environment that has supported us is to survive.

As temperatures rise, heat-reflecting snow and ice become water, which absorbs 90% of greenhouse gas (GHG) heat and creates water vapor. Warming the oceans increases their volume, which brings coastal flooding. Nevertheless, Florida's Gov. Rick Scott has told state employees to avoid discussing climate change, and Miami is launching a building boom despite street flooding from increasingly higher tides.

The loss of snow and ice exposes land, which, as it warms, produces even more water vapor, which brings heavier rains and stronger thunderstorms and tornadoes. And a warming planet will experience a decrease of snowfall, which will reduce the mountain runoff needed to replenish reservoirs that store precious water for agricultural, industrial and personal use.

As the glaciers and land-based ice in Antarctica and Greenland melt, rising sea levels will destroy coastal cities, create millions of desperate refugees and cause civil unrest. The insurance industry understands this, and has already begun to adjust its rates.

http://www.insurancejournal.com/news/national/2016/04/12/405089.htm
https://www.facebook.com/climatereality/videos/1133593866707256/
http://www.newsweek.com/climate-change-high-environmental-records-487031

For eons, Nature has relied on two primary methods to capture carbon dioxide. The first is photosynthesis by forests, crops and ocean plants that range from huge (kelp) to tiny (phytoplankton), but we are clear-cutting forests equal in area to West Virginia every year while simultaneously polluting the oceans. The second also involves the oceans, which can absorb huge amounts of carbon dioxide.

However, adding CO_2 to water creates carbonic acid, which impedes the formation of the calcium carbonate shells of crabs, shrimp, lobsters, oysters, scallops, and most importantly, tiny organisms like the phytoplankton that comprise the foundation of the ocean food chain.

We have convincing evidence that the concentration of CO_2 and other greenhouse gases will, within a few decades, equal those that caused the Permian extinction that occurred 250 million years ago - *when more than 90% of all oceanic species died* due to massive eruptions of CO_2 and methane, primarily in Siberia.

Because those conditions developed over hundreds of thousands of years, some oceanic organisms had enough time to evolve, but our anthropogenic (human-caused) Climate Change, *being much more rapid*, will leave insufficient time for many organisms to evolve.

Like it or not, the problems we face are the direct result of our creating 1.8 trillion tons of Industrial Age CO_2, to which we are adding 30 billion tons per year. However, only 1/3 of that 1.8 trillion tons has dissolved in our seas, and as the remainder is absorbed, our oceans will become even more acidic (less alkaline) and increasingly hostile to life.

For millions of years, our oceans have been slightly basic, having an average pH of 8.2. (7.0 is neutral, being neither acid nor basic.) However, in the last 250 years, our excesses of CO2 have lowered ocean pH from 8.2 to 8.1. That might seem trivial, but because the pH scale is logarithmic, not linear, this represents a *much larger increase in acidity,* and a pH of 8.0 or 7.9 will mean death to many species, including phytoplankton, *and near-death to the oceans that provide 15 to 20% of our protein and 50% of the oxygen that we require.*

Even if we stop burning carbon today, we will still have almost 1.2 trillion tons of *excess, man-made* CO2 in our atmosphere to deal with. It is no exaggeration to say that we only have a few *years, not decades,* to prevent the next 0.1 drop in pH.

From *Ocean Scientists for Informed Policy:* "It is not up for debate: It is a cold, hard fact that both climate change and ocean deoxygenation are happening."

Oceans are Acidifying Fast ....

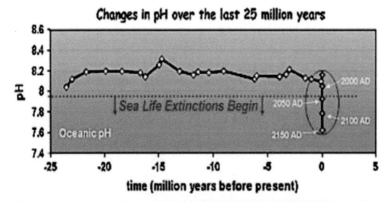

It is happening now, at a rate and to a level not experienced by marine organisms for ~20MY

Horrifying Study Finds that the Ocean is on its Way to Suffocating by 2030 - by A. Haro - *The Inertia*

According to Matt Long, an oceanographer at the National. Center for Atmos. Research, if we continue on the road we are on, the ocean could begin to suffocate in 15 years.

http://www.newsweek.com/pacific-ocean-deoxygenation-2030-climate-change-454157

http://www.whitehouse.gov/blog/2012/03/13/study-finds-ocean-acidification-rate-highest-300-million-years-co2-culprit

Since 1980, we have melted 80% of the Arctic's ice, and in 2014, scientists at California's Jet Propulsion Laboratory, who monitor the rate of arctic melting, reported that at least 50 cubic miles of the Greenland ice sheet melted during *just 2013*.

http://www.smh.com.au/environment/weather/the-north-pole-is-an-insane-20c-warmer-than-normal-as-winter-descends-20161117-gss3bg.html

In early April, 2017, the Coast Guard's International Ice Patrol**,** which tracks icebergs, sighted 450, far more than the historical average of 83 in the same area at that time of year.

As the Arctic warms, the tree line is slowly moving north, as are robins, black bears and a host of "southern" insects. I have seen these changes and many more.

From 1967 to 2008, I spent parts of 38 summers "bush flying" all across northern Canada and Alaska. There, winters are now at least five weeks shorter than they were just 50 years ago, and the shrinking icepack is leaving polar bears insufficient time to fatten up on seals, with many bears coming off of the springtime ice severely underweight. Some are drowning, having become too weak to survive what was once, for a healthy polar bear, an easy 100-mile swim to shore.

Once ashore, these weakened bears face a new hazard: Grizzly bears are expanding their range, and even a healthy, mature polar bear is no match for a grizzly bear.

Now, with NASA and NOAA reporting that *2016 was, globally, the hottest year ever recorded,* and with arctic temperatures running as high as 16 degrees F (9 degrees C) above normal, what hope is there for these magnificent animals – and for many other species that are not as photogenic or obvious?

PHOTO: KERSTIN LANGENBERGER/FACEBOOK

The North pole was <u>36 degrees F above normal</u> when winter 2016 began.

http://www.smh.com.au/environment/weather/the-north-pole-is-an-insane-20c-warmer-than-normal-as-winter-descends-20161117-gss3bg.html

In Oregon, Washington and British Colombia, oyster farmers have begun to add lime to the ocean water that fills their tanks to counter its acidity. And according to the World Wildlife Fund and the Zoological Society of London, overfishing between 1970 and 2014 has reduced the number of fish and other ocean species by **50%**, with tuna and mackerel down by 74%. In addition, several new studies show that even current levels of oceanic CO2 can even "intoxicate" fish, which can impact their ability to survive.

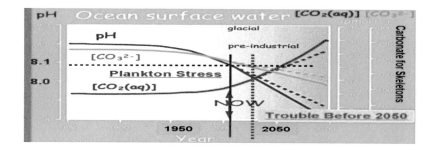

The year scale in the above image ranges from 1850 to 2100, and the dark upper line shows decreasing pH - increasing acidity. The line below it reveals the decrease in carbonate available for making shells, and "NOW" is 2014. We will be farther down the dark blue line when most people read this book.

In 2014, Canadian scientists discovered that the volume of arctic phytoplankton had dropped an alarming 40% since 1950, and since then, it has continued to drop by one percent per year.

Why should we care about these tiny plants? Because phytoplankton provide the base of the ocean food pyramid that sustains most oceanic life, and no phytoplankton will eventually mean "no fish." In addition, as previously noted, phytoplankton produce 50% of our oxygen and consume most of the carbon-dioxide we produce by using carbonates to build their shells.

When they die, their shells accumulate on the ocean floor, eventually becoming limestone – the end result of the most effective carbon sequestration process on the planet. That process can sequester a billion tons of carbon dioxide per year, which sounds impressive, but, as noted earlier, we are emitting *30 billion tons* of CO2 every year – and since prehistoric times, the amount of oxygen in our atmosphere has declined by a third, almost entirely due to deforestation and the decrease in phytoplankton.

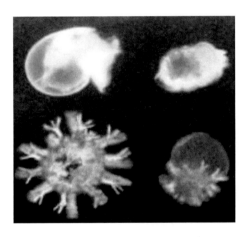

Healthy North Sea larvae left side.
Impaired larvae right side. Image - AAAS Science
http://m.phys.org/news/2015-07-ocean-acidification-phytoplankton.html

As Elizabeth Kolbert noted in *The Sixth Extinction*, "Australia's Great Barrier Reef is already 50% **dead,** and by 2050, shellfish calcification (and survival) in most oceans will have become impossible... New data finds that the rate of human-caused carbon dioxide emissions is greater than the rate of the carbon dioxide emissions from volcanic activity that marked the great extinction 250 million years ago **when the world lost 90% of all species."**

http://www.truth-out.org/news/item/38641-great-barrier-reef-suffered-worst-coral-die-off-on-record-in-2016-new-study
http://www.washingtonpost.com/news/energy-environment/wp/2015/10/08/scientists-say-a-dramatic-worldwide-coral-bleaching-event-is-now-underway/

Even if we find a way to emit less CO_2 than is being absorbed, our oceans will continue to acidify because the extra CO_2 we have already created will persist in our atmosphere for hundreds of years, and in the oceans for tens of thousands of years, which is why we must develop some form of corrective geo-engineering. Doing that will require huge amounts of *CO_2-free,* non-polluting nuclear power. Nothing else will do. In fact, reducing acidification must become a worldwide priority if we are to avoid a life-changing oceanic and humankind disaster.

Extinctions of sea life are certain if we do nothing.

Barbara Ward – "We cannot cheat on DNA. We cannot get around photosynthesis. We cannot say I am not going to give a damn about phytoplankton. All of these mechanisms provide the preconditions of our planetary life. To say we do not care is to say that we choose death."

Three potential remedies

by Dr. Alex Cannara

1. "Mimic the natural carbon sequestration process of the oceans: Use CO_2-free, highly efficient nuclear energy to heat limestone or dolomite, which releases lime – calcium oxide and magnesium oxide - which we distribute in the oceans to neutralize the carbonic acid. The CO2 that is produced when limestone is heated would be sequestered in porous basalt, with which it chemically combines. Refining enough lime from limestone will require about 900, one -Gigawatt nuclear plants – and that's only enough to neutralize our present emissions.

[If we had adopted the Atomic Energy Commission's 1962 recommendation to expand nuclear power, we'd already have those nuclear plants, we'd have created less CO2, and we'd have saved MILLIONS of lives that have been lost due to carbon-related pollution. GAE]

2. "Spread finely ground basalt into the oceans. Basalt, as produced by volcanoes, is "CO2-hungry," so it would remove carbon dioxide from the oceans. Lime and basalt, being basic, would assist shell formation by neutralizing the carbonic acid. Volcanic ash, which is primarily powdered basalt, can also be used to improve soil quality, so scattering "powdered" basalt across farm fields could help remove the excess CO2 from our atmosphere.

http://onlinelibrary.wiley.com/doi/10.1002/rog.20004/full

"Our current anthropogenic carbon dump rate is about 33.4 gigatons of CO_2/year. Each ton of powdered basalt can "fix" about 0.2 tons of carbon (0.73 ton CO_2), so we'll need to mine, grind, and disperse about 46 billion tons of basalt powder/yr to keep up with our current CO_2 dump rate (about the total amount of sand & gravel now mined/yr). At 100 kWhr/ton, the power needed to convert that much rock to powder would require the electrical output of 500, 1 GW_e nuclear reactors.

However, basalt contains many minerals, some of which might be harmful to sea life, so basalt might have to yield to lime, which is as natural as the organisms that incorporate it in their carbonate shells and skeletons. In any case, marine biologists should oversee these actions and the production of the materials.

"For this to work on land, fields should be warm, watered, regularly tilled and biologically active. The world's 400 million acres of rice fields seem to fit that bill. Land currently devoted to corn and soybean production would probably also be suitable.

"This approach is more affordable than scenarios that invoke electrochemistry or the calcination of limestone. In addition, this approach would appeal to countries that want to assist agricultural productivity, and it's probably more acceptable than "seeding" the oceans.

https://zenodo.org/record/12863/files/Gislason_et_al._GHGT-12_2.14.pdf

3. "Pump water and CO2 from the air into the basalt that underlies huge areas of the globe. The volcanic basalt, being basic, will combine with the carbonic acid to LOCK UP the CO2. This is not the same as pumping compressed CO2 down a hole and hoping it stays there.

"Iceland studies reveal that up about 150 pounds (70 kg) of CO2 can be stored in just one cubic meter of basalt, and if we could also apply this process to the basalt in ocean ridges, we could sequester the 5,000 Gigatons (Gt) of CO2 created by burning all of the fossil fuel on Earth. If this were done worldwide, it could drastically shorten the timescale of carbon trapping. Instead of taking centuries, CO2 trapping via basalt carbonation could be completed within a few decades, <u>but it will require huge amounts of carbon dioxide-free electrical power.</u>"

https://www.sciencedaily.com/releases/2016/11/161118105540.htm (Supercritical CO2 Reactivity with Basalts.)

To summarize: Our planet's ocean life can sequester a billion tons of CO2 per year by making shells, skeletons, limestone, etc. However, the 1/3 of the 1.8 - 2 trillion tons that the ocean has absorbed has already lowered ocean pH close to extinction levels for many organisms.

<u>Ocean warming has worsened the threat, and *2050, not 2100*, is the key oceanic end-of-life date – and this doesn't include the warming caused by methane released from thawing permafrost and sub-sea methane hydrates. Therefore, getting atmospheric CO2 levels down to 350 is probably meaningless, *if we don't protect ocean chemistry.*</u>

We must also electrify cement making (which requires huge amounts of energy), with electricity created by nuclear power, then sequester the CO2 released during the process in basalt and use the lime to help restore the ocean.

What we have been doing is like "taking a one-week fling, and in the process, contracting a horrible disease."
Bill McKibben

Chapter 2
The Lie

"No science is immune to the infection of politics and the corruption of power." Jacob Bronowski

In 1928, Hermann Muller, the originator of the **L**inear **N**o **T**hreshold (**LNT**) theory, exposed fruit flies to at least 2,750 milliSieverts (mSv) of radiation in just 3 1/2 minutes, which, of course, caused mutations. (Radiation **dose,** which we measure in Sieverts, is the biologically effective energy transferred to body to body tissue by radiation.)

Then, although the dose he used is comparable to 1,000 mammograms, he called it a **low** dose, although it is extremely **high**. (Even Japanese atomic bomb survivors didn't receive such a large dose.)

Muller then extrapolated his results down to ZERO mSv despite contrary evidence, and continued to promote his theory into the fifties and sixties, perhaps because he wanted to heighten fear of fallout from atmospheric testing of nuclear bombs. Muller's theory argued that there is **no safe level** for radiation and claimed that even the smallest amounts of radiation are **cumulative.**

Muller knew that his results were disputed, as did several of his colleagues, one being a meticulous researcher named Ernst Caspari, whose work Muller had repeatedly praised. (We learned all of this after Muller's correspondence became public late in the 20th century.)

In the fifties, no one knew that our cells can *easily* repair DNA damage, whether caused by radiation or oxidation, so we accepted Muller's theory.

Muller's LNT theory also asserts that, *even at low dose rates over long times,* the risk is proportionate to the dose, and it ignores our adaptive response mechanisms, which were unknown at that time.

Muller's theory is analogous to the earth-centered solar system that everyone "knew" was true for thousands of years, and it's regrettable that so many still believe it. From its beginning, the LNT theory was based on fraud and perpetuated by fear.

Muller received the 1946 Nobel prize for his 1926 discovery of X-ray mutation of fruit flies.

...these principles have been extended to total doses as low as 400 r, and rates as low as 0.01 r per minute, with gamma rays. They leave, we believe, no escape from the conclusion that there is no threshold dose, and that the individual mutations result from individual "hits", producing genetic effects ..."

Excerpt from Muller's Nobel acceptance speech.

Expert toxicologist Edward Calabrese studies dose-response effects.

Professor, University of Massachusetts, Amherst

B.S., Bridgewater State, 1968
M.A., University of Massachusetts Amherst, 1972
Ph.D., University of Massachusetts Amherst, 1973

2009 Marie Curie Prize
CV = 145 pages

So why wasn't Muller truthful? In a radio interview on IEEE SPECTRUM's "Techwise Conversations," Dr. Calabrese explained it this way:

"Ernst Caspari and Kurt Stern were colleagues, and Muller was a consultant to Stern. Muller provided the fruit fly strain that Stern and his coworkers used. Stern and Muller thought there was a linear dose-response relationship <u>even at low doses</u>....

"In the chronic study, which was done far better in terms of research methodology than an earlier study, they found that the linear relationship was *not* supported, [flawed] and what they observed would be supportive of a threshold dose-response relationship. This created a conflict—not for the actual researchers like Caspari, but for his boss, Kurt Stern, who tried to convince Caspari that his study didn't support the linear model because his control group values were artificially high.

"So Caspari... got lots of unpublished findings from Muller, and put together a case that his boss was wrong. Ultimately, he got Stern to accept his findings that <u>supported the threshold dose response</u>. [Which meant that there was a threshold below which low doses of radiation were safe.]

"They sent Caspari's paper to Muller on Nov. 6, 1946. On Nov.12 he [Muller] wrote to Stern indicating that he went over the paper, and <u>he saw that the results were contrary to what he thought would have happened,</u> that he couldn't challenge the paper because Caspari was an excellent researcher, that they needed to replicate this, and that this was a significant challenge to a linear dose response because this study was the best study to date, and it was looking at the lowest dose rate that had ever been used in such a study.

"A month later, Muller went to Stockholm to accept his Nobel Prize, and in his speech, he tells the scientists, dignitaries, press... that one can no longer accept any consideration of a threshold model, that all you can really accept is the linear dose-response model. ...Yet Muller had actually seen the results of a study that he was a consultant on, that was the best in showing <u>no support for the linear model</u> - but support for a threshold model.

"He had the audacity to actually go in front of all these dignitaries and mislead the audience. He could have said, 'This is a critical area, and we need to do more research to try to figure this out.' It would have been intellectually honest and the appropriate thing to say, but that's not what he says. He tries to actually mislead the audience by saying there's not even a remote possibility that this alternative exists, and yet he has seen it." http://tinyurl.com/4xqwzjc

Because Muller had strongly (and quite appropriately) opposed the atmospheric testing of nuclear weapons, and because he wanted to persuade Congress and the American public to oppose the expansion of nuclear energy, he seems to have concluded that the end justified the means, even if it compromised his integrity.

http://www.science20.com/news_articles/national_academy_sciences_misled_world_when_adopting_radiation_exposure_guidelines-118411

http://www.sciencedaily.com/releases/2011/09/110920163320.htm

Calabrese published the story of Muller's deceptions.

How the US National Academy of Sciences misled the world community on cancer risk assessment: new findings challenge historical foundations of the linear dose response

Edward J. Callabrese Arch Toxicol 2013

1. Muller's Nobel lecture promoted LNT, though he knew of contrary evidence.
2. Colleague Stern covered up Muller's deceptions.
3. Led to National Academy 1956 adoption of LNT.

In November, 2014, **Dr. John Boice**, president of the National Council on Radiation Protection, stated, "...the reason they were concerned about the risk of radiation doses all the way to zero was because they used a theory for genetic effects that assumed that even a single hit on a single cell could cause a mutation, and they did not believe there was any such thing as a beneficial mutation.

When the LNT model was adopted by the National Academy of Sciences in 1956, its summary stated: *"Even small amounts of radiation have the power to injure."* The report, which was published in the New York Times, quickly inflated the fear of radiation, even extremely low levels.

Linear No-threshold Theory (LNT) was proclaimed by that 1956 committee.

No minimum

There is no minimum amount of radiation which must be exceeded before mutations occur. Any amount, however small, that reaches the reproductive cells can cause a correspondingly small number of mutations. The more radiation, the more mutations.

Cumulative harm

The harm is cumulative. The genetic damage done by radiation builds up as the radiation is received, and depends on the total accumulated gonad dose received by people from their own conception to the conception of their last child.

However, "de-classified" letters between some of the members of the National Academy of Science committee indicate that the reason for adopting the LNT model was *not* that small amounts of radiation might be dangerous, but that Muller's deception (and possibly self-interest), had trumped science - one individual writing:

"I have a hard time keeping a straight face when there is talk about genetic deaths and the dangers of irradiation. Let us be honest—we are both interested in genetics research, and for the sake of it, we are willing to stretch a point when necessary... Now that the business of genetic effects of atomic energy has produced a public scare and a consequent interest in and recognition of importance of genetics. This is good, since it may lead to the government giving more money for genetic research."

In 2015, while reading Dr. Siddhartha Mukherjee's *The Emperor of All Maladies*, an eloquently written, meticulous, Pulitzer Prize winner about our long battle with cancer, I came upon the following passage:

"In 1928, Dr. Hermann Muller, one of Dr. Thomas Morgan's students, discovered that X-rays could increase the rate of mutations in fruit flies..." [Morgan, by studying an enormous number of fruit flies, had discovered that the altered genes and mutations could be carried from one generation to the next.]

"Had Morgan and Muller cooperated, they might have uncovered the link between mutations and malignancy. But they became bitter rivals.... Morgan refused to give Muller recognition for his theory of mutagenesis...

"Muller, in turn, was sensitive and paranoid; he felt that Morgan had stolen his ideas and taken an undue share of the credit. In 1933, having moved his lab to Texas, Muller walked into a nearby woods and swallowed a roll of sleeping pills in an attempt at suicide. He survived, but was haunted by anxiety and depression.

In 1933, Morgan received the Nobel Prize in Physiology or Medicine for his work on fruit fly genetics." Knowing this, I wonder if Muller's need for recognition and his resentment of Morgan might have caused him to hide the work of Ernst Caspari and others because it would have jeopardized his "fifteen minutes of fame."

Muller received his Nobel Prize in 1946, but his deception has promoted the fear of all forms of radiation, however feeble. In addition, it has caused the deaths of millions and accelerated Climate Change by stunting the growth of nuclear power, which has required us to burn coal, oil and natural gas.

https://www.21stcenturysciencetech.com/articles/nuclear.html

http://radiationeffects.org/ and http://www.x-lnt.org/

http://www.aiva.ca/Dobrzynski_L_etal_Dose-Response_2015.pdf

https://www.youtube.com/watch?v=xhkBLhw-8pk&feature=youtu.be

http://atomicinsights.com/atomic-show-224-dr-john-boice-ncrp/

http://www.ncbi.nlm.nih.gov/pmc/articles/PMC2663584/

US President John Kennedy said:

For the **great enemy of the truth** is very often not the lie—deliberate, contrived, and dishonest—but the myth—persistent, persuasive, and unrealistic. Too often we hold fast to the clichés of our forebears. We subject all facts to a prefabricated set of interpretations. We enjoy the comfort of opinion without the discomfort of thought.

"To overturn orthodoxy is no easier in science than in philosophy or religion…" Ruth Hubbard

Chapter 3

A Little Nuclear History
Beer and bananas
When Radiation Is Safe and When It Isn't

By 1969, the United States had a new, super-safe, highly efficient **Molten Salt Reactor** (MSR). Fueled by uranium dissolved in a very hot, liquid salt, the MSR had performance and safety advantages over water-cooled, uranium-powered, <u>solid-fuel</u> **Light Water Reactors** (LWRs) – hereafter also called "conventional" reactors.

LWRs are cooled with normal (light) water, a term used to distinguish them from reactors that are cooled with "heavy" water – deuterium. LWRs use pellets that contain 3.5% to 5% U-235, with the remainder being U-238 (from which U-235 is derived), but deuterium-cooled reactors can utilize un-enriched U-238. Most nuclear reactors in use today are LWRs.

During the sixties, Alvin Weinberg, the Director of Oak Ridge National Laboratories, proved the superiority of MSRs in hundreds of tests during 22,000 hours of operation, but due to the success of conventional reactors in submarines, Admiral Hyman Rickover's water-cooled reactors became the choice for commercial power production. Weinberg, who protested that MSRs were safer and more efficient, was fired, and the MSR program was terminated.

There was a second reason: The Cold War was heating up, and the uranium-plutonium fuel cycle of LWRs could be adapted for making bombs, but making a nuclear weapon with MSR technology was, and still is, difficult and dangerous.

The Atomic Energy Commission knew that Molten Salt Reactors could generate abundant, low cost, 24/7 electricity while breeding their own fuel from U238 or Thorium – and that Thorium would create far less waste than conventional reactors.

Had we switched to MSRs in the sixties, we would have eliminated much of the CO2 that created Climate Change and drastically reduced the toxic emissions from burning coal that have caused medical expenses that cost billions of dollars.

From the April, 2013 *Scientific American:*

"**James Hansen**, former head of the NASA Goddard Institute for Space Studies, has said that just our *partial* reliance on carbon-free nuclear power since 1971 has saved 1.8 million lives that would have been lost due to fossil fuel pollution. By contrast, we assess that large-scale expansion of natural gas use would not mitigate the climate change problem and would cause *more* deaths than expansion of nuclear power."

However, because we rejected MSRs, almost all of the electricity we have generated with nuclear power has been produced by LWRs - high pressure, water-cooled reactors that are fueled with uranium pellets – a workable-but-complex process. Unfortunately, according to Michael Mayfield, head of the Office of Advanced Reactors at the Nuclear Regulatory Commission, the NRC is "unfamiliar with most, new small-reactor technology, [including MSRs] and has no proven process to certify one." (2010) **THAT MUST CHANGE!**

In 2013, the U. S. Energy Information Administration predicted that world energy use will increase 56% by 2040 - and most of that increase will come from burning carbon-based fuels, which will add even more CO2 to our already damaged biosphere.

Why not replace CO2-producing power plants with GREEN nuclear power? Is it really as hazardous as some claim?

The largest obstacle to expanding nuclear power is the fear caused by misinformation about radiation safety, so let's begin with a question intended for seniors like me: "Do you still have your toes?"

This foolish sounding question refers to a machine that, during the thirties and forties, stood near the entrance of every up-to-date shoe store in America. Called the ADRIAN shoe-fitting machine, it was ballyhooed as the perfect way to see if one's shoes fit properly.

Attractive ads with photos of the marvellous machine proclaimed, "Now, at last, you can be certain that your children's foot health is not being jeopardized by improperly fitting shoes. If your children need new shoes, don't buy their shoes blindly. Come in and try our new ADRIAN Fluoroscopic Shoe Fitting machine. Use the new, scientific method of shoe fitting that careful parents prefer."

The customers, usually children, inserted their feet into an opening while their parents watched the image in two viewing ports. Unattended children would often repeatedly switch sides to watch their siblings' toes wiggle. It was fun, and no-one gave a thought to X-ray exposure.

But despite these frequent, fairly high exposures, no malignancies or other damage to the feet of foot-radiating junkies like me were ever reported.

Now, as I travel the country with my presentations on nuclear power and radiation safety, I always ask the seniors in my audiences, all of whom instantly recognize the machine, if they still have their toes.

During 2016, I queried some 1,000 seniors, but I never found any evidence of damage. However, my tale of the shoe-fitting machine always brings laughter and an opportunity to talk about the Merchants of Fear whose hype created the 20th century word: ***radiophobia.***

Dr. Alex Cannara

"We've accepted for decades that millions of people are allowed to be killed by combustion pollution and mass-produced weapons. We've accepted for at least 100 years that the planet's climate and oceans can be allowed to be changed for the worse because of our love of combustion. We even accept poverty and all its ill effects, simply due to our general inaction. But the safest form of energy production, nuclear power, is foolishly married to fear of nuclear weapons."

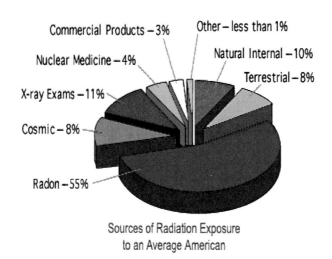

Sources of Radiation Exposure to an Average American

Radiation from nuclear power is just a tiny part of the 1% listed above as "other."

Most people do not realize that we are bathed in radiation for our entire lives – about 2/3 from cosmic radiation and elements like radon, and the rest from elements *within us* plus from medical use and consumer products like smoke detectors. We all have some 4,400 beta/gamma decays **per second** throughout our bodies for life, largely from the Potassium 40 (K40) in foods like bananas and potatoes. (Eating bananas is more "dangerous" than living beside a nuclear power plant for an entire year.)

"Fear and paranoia are the two most common forms of radiation sickness." **Mike Conley**

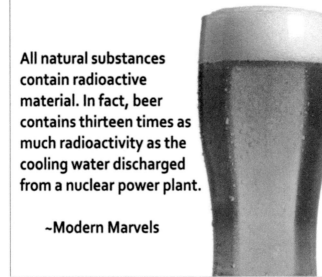

> All natural substances contain radioactive material. In fact, beer contains thirteen times as much radioactivity as the cooling water discharged from a nuclear power plant.
>
> ~Modern Marvels

http://www.ncbi.nlm.nih.gov/pmc/articles/PMC4036393/
http://www.gizmodo.com.au/2015/07/your-fear-of-radiation-is-irrational/

Our ancestral life forms thrived during times when radiation levels were far *higher* than they are today. As a consequence, they evolved some very effective ways to repair the damage to the DNA in our cells caused by radiation *and oxidation*, which is why we are told to favor anti-oxidants like grapes and greens. (DNA is "short" for deoxyribonucleic acid, a complex, spiral, chain-like molecule that contains our genetic codes.)

However, even the highest natural background radiation rate is insignificant when compared to the damage caused by our internal chemistry. <u>DNA bond breaks caused by oxidation and toxins occur at least 1,000 times more frequently than breaks caused by background radiation</u>.

If people understood that "...we have billions of cells that die every day and must be replaced, they will be better able to accept the fact that our bodies have efficient repair mechanisms that can handle low level radiation". (Adults have about 37 trillion cells.) **SCIENCE** magazine, March, 2015.

Nobel Prize Awarded to Lindahl, Modrich and Sancar for DNA Studies

NYT 10-7-2015

"Each cell contains a coiled mass of DNA that carries the thousands of genetic instructions that we need to run our bodies. These strands of DNA undergo thousands of spontaneous changes every day, and DNA copying for cell division and multiplication, *which happens in the body millions of times daily,* also introduces defects.

"DNA can be damaged by ultraviolet light from the sun, industrial pollutants and natural toxins like cigarette smoke. *What fights pandemonium are our DNA repair mechanisms.*

"In the 70s, Dr. Lindahl defied orthodoxy about DNA stability by discovering a molecular system that counteracts DNA collapse, and Dr. Sancar mapped out how cells repair DNA damage from UV light. People born with defects in this system, when exposed to sunlight, develop skin cancer, and Dr. Modrich showed how our cellular machinery repairs errors that arise during DNA replication, thereby reducing the frequency of error by about 1,000."

http://youtu.be/UzXcq2h0VCk?t=7m10s

All radioactive elements "decay" by emitting charged particles and/or electromagnetic waves, eventually becoming stable, non-radioactive elements. Their "half-life" is the time needed for half of the atoms in a given mass to decay. At the end of that time, the "level" of the radiation will be 50% of its original value. For the potassium-40 in our bananas and bodies, it is 1.2 billion years. For the Americium-241 in our smoke detectors, it's 432 years, and for Iodine-131, it's 8 days. Elements with long half-lives decay so slowly that they present very little risk, but elements with short half-lives can be very hazardous.

Radioactivity is measured by the number of decays per second. One decay per second is one Becquerel (Bq). One banana produces about 15 Bq from its potassium-40, but smoke detectors emit 30,000, so when nuclear power critics fuss about 64,000 Bq entering the ocean at Fukushima, remember that 64,000 Bq is equal to 14 seconds of potassium radiation activity that occurs inside our bodies every day. (The radioactivity of seawater is 12,000 Bq per cubic meter.)

However, focusing on Becquerels without factoring in the *energy* absorbed by the body is pointless: You can throw a bullet or you can shoot one, but only one will cause harm.

Fortunately, radiation is easy to detect. A single emission of just one atom per second (1 Bq) will trigger a click in any decent detector – and please recall that our own internal Potassium-40 "clicks" 4,400 times per second for life.

Dr. Timothy Maloney

"The word 'radioactivity' doesn't account for the *energy* propelling the emissions, so quoting large Becquerel counts says nothing about risk. However, big numbers can frighten uninformed people, and in building their case against nuclear power, many environmentalists have been doing just that."

https://www.patreon.com/posts/2901438?login=cannara%40sbcglobal.net or tinyurl.com/q9d7neq.

As noted earlier, radiation *dose*, which we measure in Sieverts, is the biologically effective energy transferred by radiation to tissue. For example, one mammogram equals 1 to 2 milliSeiverts (mSv), and one dental X-ray (0.001 mSv) is nowhere near enough to cause concern.

Let's now consider the normal *background radiation* that accompanies us throughout the years.

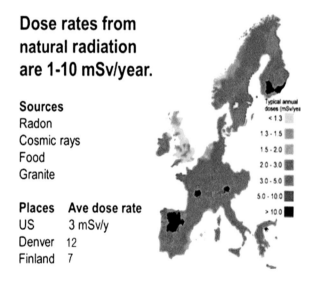

Natural "background" radiation *dose rates* vary widely, averaging 1 mSv/year in Britain, 3 in the US, 7 in Finland, 10 in Spain, 12 in Denver and up to 300 in Kerala, India. Given these statistics, one might expect cancer rates in Finland and Spain to be higher than in Britain, but Britain has higher rates of cancer than both Spain and Finland despite LNT dogma.

Dose Rates and Health

A massive, single, whole-body radiation dose - as at Hiroshima and Nagasaki severely injures blood cell production and the digestive and nervous systems.

A *single* 5,000 mSv dose is usually fatal, <u>but if it is spread over a lifetime it is harmless because at low dose rates, cells recover.</u> (Consume a cup of salt in one sitting, and you will probably die, but do it over a few months or more and it won't be a problem.)

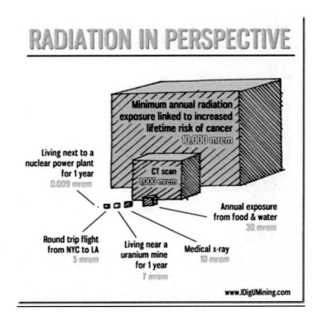

Why radiation is safe below 100 mSv/y.

In 1945, the U S exploded two atomic bombs over Japan, killing 200,000 people. Since then, 93,000 survivors have been studied for health effects. In 55 years, 10,423 of those survivors died from cancer, which is 573 (5%) more than the number of deaths expected by comparison with unexposed residents.

However, <u>no excess cancer deaths have been observed in those who received radiation doses below 100 mSv.</u> In fact, <u>Japanese A-bomb *survivors* have been outliving their unexposed peers.</u>

Subsequent studies by the United Nations Scientific Committee on the Effects of Atomic Radiation (UNSCEAR) have proved that below 100mSv, <u>which is well above normal background radiation levels, it is not possible to find any cancer excesses.</u>

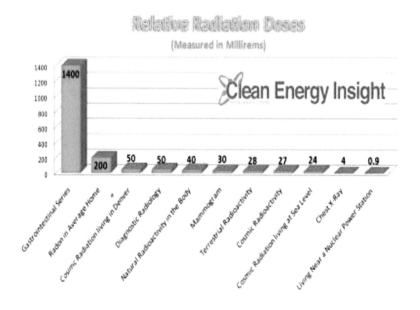

Living near a coal-fired power plant gets a 90, which is 100 X more than nuclear. Coal-fired power plants release more radiation than nuclear power plants.

Chapter 4

DNA and Hormesis

When <u>Low Level</u> Radiation Can Even Be *Good* for You!

Kerala

Near the end of the 19th century, researchers at the Massachusetts Institute of Technology (MIT) discovered that DNA strands can break and repair about **10,000** times per day <u>per cell, (This is not a typo!</u>), and that a 100 mSv per year dose increases the number of breaks by only 12 per day.

<u>In addition, the overwhelming majority of DNA breaks are caused by **ionized oxygen atoms** from the normal metabolism that constantly occurs within our cells.</u> And because DNA is a *double* helix, the duplicate information in the other strand lets enzymes easily repair single strand breaks. In fact, <u>our cells have been repairing DNA breaks since *forever*, and they have become extremely good at it.</u> http://youtu.be/UzXcq2h0VCk?t=7m10s

DNA strand breaks occur frequently.
Ionized oxygen molecules from metabolism are the principal causes.

 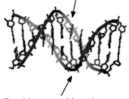

Single strand breaks occur 10,000 times per day per cell.

Double strand breaks occur 10 times per day per cell.

100 mSv/y radiation adds 12 per day.

100 mSv/y radiation adds 1 per year.

DNA is repaired.

Special enzyme DNA ligase encircles the double helix to repair a broken strand of DNA.

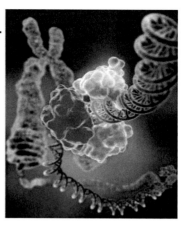

Adaptive response:
The Vaccination effect called Hormesis

Dr. Alex Cannara explains it this way:

"Radiation from unstable isotopes is always decreasing. That's what the "half-life" for an isotope expresses. Going back in time is going back to much higher radiation environments -- 8 times more for U-235 when photosynthesis began to make oxygen common in air, and oxidation made elements like Uranium soluble in water. Living things were, back then, even more intimately in contact with radioactive isotopes.

"So how did life survive higher radiation, and how did it survive the increasing oxygen atmosphere, which corrodes life's hydrocarbons into CO2 and water?

"The answer is simple: Nature evolved repair mechanisms. Each cell repairs proteins or digests badly malformed cells. Each cell repairs genetic material before it's copied for reproduction.

"A DNA or protein molecule, or one of the many repair molecules in our cells, doesn't know if a bond has been broken by an oxidizing radical, an alpha particle, or a microbial secretion. Our cellular-repair systems have evolved to fix defects regardless of cause. Thus, Nature has, for billions of years, been able to deal with chemical and radiation threats. Today, chemical threats have increased because of industry, but radiation threats have decreased.

"Therefore, we should not be surprised by the absence of radiation deaths at Fukushima and the small death rates in and around Chernobyl."

We have also learned that low dose irradiation of the torso is an effective treatment for malignant lymphomas, and several new studies have shown that lung cancer deaths decreased with slightly increased radon levels in homes.

Fear of radon has been hyped by EPA's devotion to the LNT theory, and their efforts have greatly assisted those who sell and install radon-related equipment, whether needed or not. (However, studies of every U S county have revealed that those with low levels of radiation actually had higher levels of lung cancer than counties with high levels – where the incidence was lower!)

The EPA recommends radon remediation when radioactivity measures **4** pico-curies per liter of air, but an average adult is naturally radioactive at about 200,000 pico-curies. If the EPA knows this, and they should, why are they concerned about such low natural radon levels?

http://www2.lbl.gov/Science-Articles/Archive/radon
https://www.epa.gov/sites/production/files/2015-05/documents/hmbuygud.pdf
http://www.mn.uio.no/fysikk/tjenester/kunnskap/straling/radon-and-lung-cancer.pdf
http://www.symmetrymagazine.org/article/this-radioactive-life
http://www2.lbl.gov/Science-Articles/Archive/radon

Here's radon, by county, blue lowest. More than 10% of homes in non-blue counties have radon exceeding EPA's warning level.

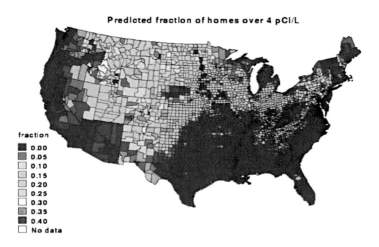

But compare the two maps. The counties with less radon have more lung cancer deaths. EPA's LNT theory is clearly wrong.

http://www2.lbl.gov/Science-Articles/Archive/radon-risk-website.html

Here are US lung cancer deaths, by county. Red counties have the highest death rates, blue lowest.

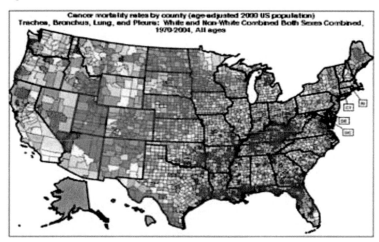

https://ratecalc.cancer.gov/ratecalc/

The S E states had the lowest radon levels, but the highest cancer rates. And a recent study of Chernobyl cleanup workers yielded similar results.

Hormesis

Surviving Chernobyl emergency workers have fewer cancers.

Dr. Zbigniew Jaworowski, MD PhD DSc, former Chairman of the United Nations Scientific Committee on the Effects of Atomic Radiation (UNSCEAR) stated:

"What is really surprising, however, is that data collected by UNSCEAR and the Forum show **15% to 30% fewer cancer deaths among the Chernobyl emergency workers** and about **5% lower solid cancer incidence** among the people in the Bryansk district (the most contaminated in Russia) in comparison with the general population. In most irradiated group of these people (mean dose of 40 mSv) the **deficit of cancer incidence was 17%**."

Because of their daily exposure to low levels of radiation, which seems to stimulate the DNA repair system, nuclear power plant workers get 1/3 fewer cancers. They also lose fewer work days to accidents as office workers.

Knowing this, it is no surprise that, when steel containing cobalt-60 was used to build Taiwan apartments, which exposed 8,000 people to an additional 400 mSv of radiation during some twenty years, cancer incidence was sharply down, not up 30% as LNT would have predicted. Instead, the residents' adaptive response to low-level radiation seems to have provided health benefits. The following chart reveals *lower* cancer rates for those who receive extra *low level* radiation vs. those who only get *background* radiation.

The US defunded low-dose radiation studies, which disprove LNT.

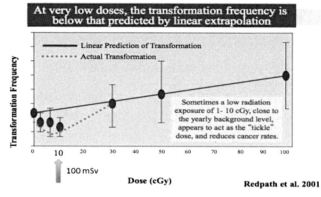

Redpath et al. 2001

In 2015, a study of bacteria grown at a dose rate 1/400 of normal background radiation yielded a surprising stress response and a <u>reduction</u> in growth. When the cells were returned to normal background radiation levels, growth rates recovered. The conclusion: <u>Insufficient</u> radiation can yield harmful results.

Given this information, it seems reasonable that radiation limits should be the same regardless of the source of the radiation. Nevertheless, nuclear plants are held to a standard 100 times higher than coal plants, which actually emit *more* radiation than nuclear power plants. Even granite buildings irradiate their occupants more than nuclear power plants.

In 2004, the **Radiation Research Society** published Tthe *Mortality Experience amongst U. S. Nuclear Workers after Chronic Low-Dose Exposure to Ionizing Radiation:*

> "Workers employed in fifteen utilities that generate nuclear power in the U. S. have been followed for up to 18 years between 1979 and 1997.
>
> "Their cumulative dose from whole body radiation has been determined from the records maintained by the facilities themselves, and also by the... Nuclear

Regulatory Commission and the Energy Department. Mortality in the cohort ... has been analyzed with respect to individual radiation doses. The cohort displays a very substantial healthy worker effect, i.e. <u>considerably lower cancer and non-cancer mortality than the general population.</u>"

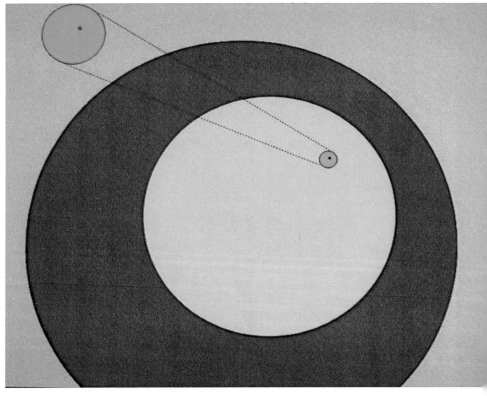

The large circle represents the dose to a tumor treated by radiotherapy;

The second largest circle represents a *recoverable* dose to healthy tissue near the tumor;

The two, small circles represent a dose with a 100% safety record;

The tiny **black dots** in the small green circles represents the limit recommended by current regulations.

In **Radiation and Health,** Hendrickson and Maillie wrote "...during radiation therapy for cancer, we've learned that chromosome damage to lymphocytes can be *reduced* by up to 50% if a small dose is given to the cells a few hours before the larger 'cancer-killing' dose is administered."
www.world-nuclear.org/info/Safety-and-Security/Radiation-and-Health/Nuclear-Radiation-and-Health-Effects/
http://go-nuclear.org/videos/item/441-positive-effects-of-low-dose-radiation-jerry-cuttler-video
https://www.youtube.com/watch?v=1rW-EwP-DNE

Kerala

In the southwest Indian state of Kerala, children under five have the lowest mortality rate in the country, and life expectancy is 74 despite background radiation rates that can range as high as 30 times the global average. (For the source of much of the following material, please visit http://bravenewclimate.com/2015/01/24/what-can-we-learn-from-kerala/.)

For thousands of years, Keralites have lived with radiation three times the level that caused the evacuation at Fukushima, where the limit was, on July, 2016, just 20 mSv/yr. In contrast, some sections of Kerala experience seventy mSv, with a few areas measuring 500 mSv - and many Keralites also eat food that is about 5 times as radioactive as food in the United States.

Despite these elevated radiation levels, the rate of cancer incidence in Kerala is much the same as the rate in greater India; which is about 1/2 that of Japan's and less than a third of the rate in Australia. As the linked article says, "Cancer experts know a great deal about the drivers of these huge differences, and radiation isn't on the list."

In Kerala, scientists have been working with a genuinely low rate of radiation exposure that mirrors what would be the case in Fukushima if the Japanese government hadn't panicked and needlessly evacuated so many thousands of people.

So, why did they? Partly from fear, but primarily because most radiation protection standards have been derived from studies of Japanese atomic bomb victims who received their dose in a very short time, and being bombed is very different from living for years with a slightly higher radiation level.

Kerala also confirms our modern knowledge of DNA repair - namely that radiation damage is not cumulative at background dose rates up to 30 times normal, and that 70 mSv over a lifetime does *nothing*. In fact, the concepts of an "annual dose" or a "cumulative dose" are more than misleading. Instead, the best available evidence is that an annual exposure to 100 mSv is comparable to a dose of zero because it doesn't exceed a person's capacity for repair.

In the past, when experts discussed these issues they couldn't consider delivery rates or DNA repair because the power and mechanisms of DNA repair were not known until long after the LNT theory was adopted. As a consequence, the suffering caused by this obsolete "science" has been immense, even causing U K radiation expert Malcolm Grimston to characterize the Fukushima evacuation as being "stark raving mad".

When the Japanese Government lifted the evacuation orders on Minamisoma City in 2015 because the radiation level had dropped to 20 mSv, city officials predicted that 80 percent of residents would not return because of their fear of radiation.

This despite the fact that the most highly irradiated areas near the plant received only 12 to 25 mSv in the first year, which is about 1/4 of the lowest dose linked to an observable increase in cancer, and at Guarapari beach in Brazil, residents often bury themselves in sand that yields <u>340 mSv</u> without ill effect.

We should be concerned about genuinely dangerous isotopes, but we shouldn't waste energy and money cleaning up minor radioactivity that doesn't do anything - as in Japan.

Despite our learning that our cells have amazing repair abilities, LNT advocates still create the radiophobia that caused the extreme evacuations at Fukushima and the flood of European abortions that followed Chernobyl. In my opinion, people who refuse to examine the evidence that negates this discredited illusion have abandoned their integrity.

Chapter 5

The Consequences of Overreaction
Alarming ALARA

"LNT was pushed through the U.N. by Russia and China in the 1950s to stop America's above-ground weapons testing. It worked, but it also caused a worldwide fear of radiation below levels that are dangerous, e.g., (0.1 Sv/yr). The radsafety people liked it because it seemed so... conservative. But it has become an ideology "ruled by hysteria and fueled by ignorance." **Dr. Kathy Reichs.**
http://www.ncbi.nlm.nih.gov/pmc/articles/PMC2663584/

Japan evacuated the black-lined area.

IAEA would recommend evacuating the red area.

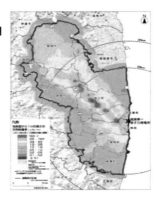

Dr. Tim Maloney: "Anyone living permanently in the green zone would only receive a dose rate equal to twice the rate in Colorado, where the cancer rate is *less* than the U S average. The dose rate in the dark red regions is 1/3 of the safety threshold set by the International commission on System of Radiological Protection in 1934. Even by today's extreme standards, this level of exposure carries no known cancer risk.

"Anxious to impress, officials and reporters donned impressive white suits and masks, which made good TV but did nothing for the child who saw the school playground being dug up by workers who were afraid of an unseen evil called *radiation*. Unfortunately, most people see their fears confirmed as fact when workers and officials dress this way. An open-necked shirt with rolled-up sleeves, a firm hand shake and a cup of tea would be a better way to reassure."

Imagine the anxiety created by clueless officials who provided useless information, as when a school official warned parents that the radiation intensity was 0.14 **micro**sieverts per hour, which was meaningless because the *normal* radiation level in some Japanese cities can be five times that high.

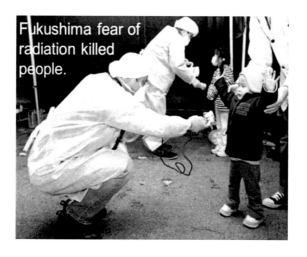

In 2012, UNSCEAR stated, "...no clinically observable effects have been reported and there is no evidence of acute radiation injury in any of the 20,115 workers who participated in Tepco's efforts to mitigate the accident at the plant."

A year later, UNSCEAR added: "Radiation exposure following the accident at Fukushima Daiichi did not cause any immediate health effects. It is unlikely [that there will be] any health effects among the general public and the vast majority of workers."
www.unis.unvienna.org/unis/en/pressrels/2013/unisinf475.html
http://www.aiva.ca/Dobrzynski_L_etal_Dose-esponse_2015.pdf

And in an April, 2014 follow-up, UNSCEAR reported that, "Overall, people in Fukushima are expected on average to receive less than 10 mSv due to the accident over their whole lifetime, compared with the 170 mSv lifetime dose from natural background radiation that most people in Japan typically receive."

Finally, eighteen months later, UNSCEAR confirmed that none of the new information accumulated after the 2013 report "materially affected the main findings in, or challenged the major assumptions of, the 2013 report." However, despite these positive reports, as of November, 2016, most of the 150,000 people who were forced to evacuate still lived in temporary housing.

Jane Orient, who practices internal medicine agreed: "The number of radiation casualties from the meltdown of the Fukushima nuclear reactors stands at zero. In Fukushima Prefecture, the casualties from radiation *terror* number more than 1,600... The U.S. is vulnerable to the same radiation terror as occurred in Japan because of using the wrong dose-response model, which is based on the linear no-threshold hypothesis (LNT), for assessing radiation health risks."

http://go-nuclear.org/radiation/item/891-fukushima-and-radiation-as-a-terror-weapon-jane-orient

The following passage is an excerpt from *Whole-body Counter Surveys of over 2700 babies and small children in and around Fukushima Prefecture from 33 to 49 months after the Fukushima accident:*

"The BABYSCAN, a whole-body counter (WBC) for small children, was developed in 2013, and units have been installed at three hospitals in Fukushima Prefecture. Between December, 2013 and March, 2015, 2702 children between the ages of 0 and 11 have been scanned, and *none* had a detectable level of cesium-137."

Positive reports like this rarely appear in our American press, which frustrates professionals like **Leslie Corrice,** a former nuclear power plant operator, environmental monitoring technician, health physics design engineer, public education coordinator and emergency planner who writes the informative and highly respected blog, *The Hiroshima Syndrome.*

In *Radiation: The No-Safe-Level Myth,* Corrice wrote,
"As long as the LNT theory is maintained, our fear of radiation will continue to damage the psyche of all humanity, restrict the therapeutic and healing effects of non-lethal doses of radiation, limit the growth of green nuclear energy, and needlessly prolong the burning of fossil fuels to produce electricity.

"In 1987, I was frustrated because it seemed like the major news outlets bent over backwards to broadcast negative nuclear reports while seemingly ignoring anything positive. "A former Press manager with a major news outlet in Cleveland, Ohio, took me aside and gave me the facts of life.

"He first explained that the Press is a money-making venture. The ratings determine advertising income; the lifeblood of the business – and the sure-fire money-makers were war, presidential elections, natural disasters and airline crashes.

"Turning to Three Mile Island, he said the ratings sky-rocketed and stayed that way for the better part of two weeks. In the years that followed, the media found that negative reports caused an increase in ratings, and positive stuff didn't. This trend slowly dwindled, but Chernobyl re-ignited the ratings impact of nuclear accident reporting and proved that broadcasting the negative was better for business....

"He added that the media might someday entirely ignore the positive and only report the negative in regard to nuclear energy, and he speculated that all it would take was one more accident. Unfortunately, he was right. Fukushima has pushed the world's Press into the journalistic dark side. My Fukushima Updates blog has lashed the Japanese Press and the world's news media outside Japan severely for primarily reporting the negative.... A recent example concerns the child thyroid study in Fukushima Prefecture during the past four years.

"On October 5, 2015, four PhDs in Japan alleged in the Tsuda Report that the Fukushima accident had spawned a thyroid cancer epidemic among the prefecture's children, which contradicted the Fukushima Univ. Medical School, Japanese Research Center for Cancer Prevention and Screening, and National Cancer Center, which all found that the detected child thyroid pre-cancerous anomalies in Fukushima Prefecture cannot be realistically linked to the accident. Regardless, the Tsuda Report's claim made major headlines in Japan, then spread to mainstream outlets outside Japan, including UPI and AP.

"Here's the problem. In December 2013, a scientific report was published on a comparison of the rate of child thyroid, pre-cancerous anomalies in Fukushima Prefecture with the rates in three prefectures hundreds of kilometers distant: Aomori, Yamanashi and Nagasaki.

"The Fukushima University medical team studying the issue had discovered that there was no prior data on child thyroid cancer rates in Japan, so there was nothing to compare the 2012 results to.

"Because of the furor caused by the original release of their findings in 2012, the team decided to take matters into their own hands and offer free testing to volunteer families in the distant prefectures. Nearly 5,000 parents took advantage of the opportunity and had their children screened.

"What was found was completely unexpected. The abnormality rates in Aomori, Yamanashi and Nagasaki Prefectures were actually *higher* than that discovered in Fukushima Prefecture, which conclusively indicated that the radiation from the Fukushima accident had no negative impact on the health of the thyroid glands in Fukushima's children. Just one Japanese Press outlet mentioned the 2013 discovery at the very end of an article about a few more children being found to have the anomalies in Fukushima....

"On the other hand, when a maverick team of four Japanese with PhDs publish a highly questionable report - full of so many holes that it should be tossed into the trash – alleging a severe cancer problem caused by the Fukushima accident, it gets major coverage inside Japan and significant coverage by the world's mainstream press!

"It is important to emphasize that the Tsuda Report fails to acknowledge the fact that Prefectures unaffected by the Fukushima accident had the *higher* anomaly rates. (Which is why the Tsuda Report is worthy of the trash heap.)

"The media might not make money off sharing the good news about Fukushima, but they are committing a moral crime against humanity by not doing it."

http://www.thenewamerican.com/world-news/asia/item/19253-fukushima-s-children-aren-t-dying

Corrice's dismay over the results of radiophobia are echoed by many professionals, one being Dr. Antone Brooks, who grew up in "fallout-drenched" St. George, Utah, which led him to study radiation at Cornell University. For an excellent, short video of the conclusions he reached, please visit https://www.youtube.com/watch?v=C0_gMpsVB-k.

And as Dr. Gunnar Walinder, an eminent Swedish radiation scientist, bluntly told UNSCEAR, "I do not hesitate to say that the LNT is the greatest scientific scandal of the 20th Century."

Alarming ALARA

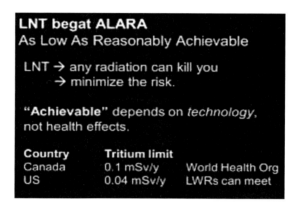

The belief that even tiny amounts of radiation can be lethal created ALARA – **A**s **L**ow **A**s **R**easonably **A**chievable – an anti-nuclear bias that has permeated our regulations for decades. However, "reasonably" is vague, and "achievable" depends on technology, not health effects.

For example, the World Health Organization has set a public exposure limit for tritium from nuclear power plants of 0.1 mSv *per year*. Canada's reactors comply with this limit, but due to ALARA, our limit is 0.04 mSv per year. Why? Because it was *achievable* - not because it is necessary.

Another example: Tritium (AKA hydrogen-3), is often used in watches and emergency exit signs, and it is also present in our food and water. Furthermore, its tiny nucleus emits a particle so slow that it can't even penetrate skin. In comparison, the potassium-40 in our omnipresent banana emits beta particles that are 230 times as energetic, but no one worries about those deadly bananas.

LNT and ALARA can easily lead to absurdities: For example, airline passengers are exposed to about 20 times more cosmic radiation than those at ground level, but despite the dire predictions of LNT, they experience no more cancer than those who don't fly. Should jets be required to fly at low altitudes, where they produce more greenhouse gases, just to satisfy ALARA – and what about the flight attendants and pilots who constantly work in higher levels of comic radiation?

As radiation detection technology improves, ALARA just increases fear.

0.6617 MeV γ radiation energy is a signature of cesium-137.

Cesium-137 from Fukushima is detectable, so *CounterPunch* complains of Bluefin tuna with 0.0000077 mSv per 7 oz serving writing...*no radiation exposure of any kind is "safe"*...

It is wasteful to spend money "protecting" people from tiny amounts of radiation. Instead, let's finance programs that help people stop smoking, which brings carcinogens like cyanide, formaldehyde, ammonia, carbon monoxide and nitrogen oxide into intimate contact with their lungs. (Smoking-related diseases kill 5 million people per year.)

Radiation exposure in reactor buildings is so low that it isn't an issue, but educating the public on the basics of environmental radiation is a very critical issue: For example, after Fukushima, lack of accurate radiation knowledge plus the media's eagerness to hype radiation issues caused a shortage of potassium iodide (KI) pills along our western coast, but no media explained that this was pointless. Pharmacies ran out, and some patients who needed KI couldn't get it, while those who needlessly took it actually raised their chances of disease because too much KI can cause thyroid malfunction.

> **Radiation is safe within limits.**
>
> - LNT and ALARA are regulation **policies**, not scientific **facts**. Replace them.
> - An evidence-based radiation safety limit would be 100 mSv/y.
>
> - *Rational regulation* is *all that is needed* to let nuclear power thrive and solve our global environmental and economic crises.

The current U. S. radiation limit for the public is less than 1 mSv/year. However, <u>50 mSv/yr is the new limit proposed by Carol Marcus and other experts in a 2015 petition that requests the NRC to increase the limits based on current knowledge.</u>

For a comparison of the consequences of accepting LNT (which led to ALARA), please see

http://www.aiva.ca/Dobrzynski_L_etal_Dose-Response_2015.pdf
http://radiationeffects.org/ and http://www.x-lnt.org/

Absurd Radiation Limits Are a Trillion Dollar Waste
Forbes magazine – 2014

"There are some easy decisions to make that will save us a trillion dollars, and they could be made soon by the Environmental Protection Agency. The EPA could raise the absurdly low radiation levels considered to be a threat to the public. <u>These limits were based upon biased and fraudulent "research" in the 40s through the 60s, when we were frightened of all things nuclear and knew almost nothing about our cells' ability to repair damage from excess radiation.</u>

"These possible regulatory changes have been triggered by the threat of nuclear terrorism, and by the unnecessary evacuation of tens of thousands of Japanese after Fukushima Daiichi, and hundreds of thousands of Russians following Chernobyl. There, the frightened authorities were following American plans created under the ALARA policy (As Low As Reasonably Achievable) that has always been misinterpreted to mean any and all radiation is dangerous - no matter at what level. It's led to our present absurdly low threat level of 25 millirem (250 *micro*Sv). "Keep in mind that radworkers can get 5,000 mrem/year and think nothing of it. We've never had problems with these levels. Emergency responders can get up to 25,000 mrem to save human lives and property. I would take 50,000 mrem just to save my cat.

"This wouldn't be bad if it didn't have really serious social and economic side-effects, like pathological fear, significant deaths during any forced evacuation, not receiving medical care that you should have, shutting down nuclear power plants to fire up fossil fuel plants, and a trillion-dollar price tag trying to clean-up to levels even Nature doesn't care about."

Chapter 6

What's so Great about Nuclear Power?
Three Mile Island, Chernobyl and Fukushima

No other technology produces energy as cheaply, safely and steadily on a large scale as nuclear power. No other energy source can match nuclear power's low environmental impact, partly because its energy density is a million times greater than that of fossil fuels – and more so for wind or solar.

As of 2013, the world's 400 + nuclear reactors generated about 15% of the world's electricity. France tops 60%. Finland, currently at 30%, is aiming for 60, and Sweden plans to add 10 reactors. Nuclear France emits about 40 grams of CO2/kwh, but Germany, the US, Japan and most industrialized nations emit 400 - 500 grams per kilowatt hour - *ten times more per kwh than nuclear France.* Compared to fossil fuel-burning wind and solar, nuclear power is a gift from the energy gods.

Nuclear power is the most effective displacer of greenhouse gases, so how can my fellow "greens," oppose CO2-free nuclear power when the environmental costs of burning carbon-based fuels are so high?

Doctor James Lovelock, a patriarch of the environmental movement, has begged his friends to drop their objection to nuclear energy: "…its worldwide use as our main source of energy poses an insignificant threat compared with the dangers of lethal heat waves and sea levels rising…. civilization is in imminent danger and has to use nuclear power, the one safe, available, energy source now or suffer the pain soon to be inflicted by an outraged planet." (*Power to Save the World* by Gwyneth. Cravens)

In May, 2014, Robert Bryce wrote in *Bloomberg View*,

"In the core of just one reactor, the power density is about 338 million watts per square meter. To equal that with wind energy, which has a power density of 1 watt per square meter, you'd need about 772 square miles of wind turbines….

"Some opponents still claim that nuclear energy is too dangerous. Debunking that argument requires only a close look at the facts about Fukushima….

"Here's the reality: The tsunami caused two deaths -- two workers who drowned at the plant.

"It was feared that radiation from the plant would contaminate large areas of Japan and even reach the U.S. That didn't happen. In 2013, the World Health Organization concluded: 'Outside of the geographical areas most affected by radiation, even within Fukushima prefecture, the predicted risks remain low and no observable increases in cancer above natural variation in baseline rates are anticipated.

"High on my list of well-intentioned dupes are those who praise science and are eager to confront Climate Change, but refuse to accept nuclear power as an essential part of carbon-reduction strategies. They dismiss new reactor designs that they don't understand, and then talk about how wind and solar power can "supply our needs."

"They are wrong, but nuclear can supply our needs when people conquer their fears, educate themselves on the safety of nuclear power - and constructively join the fray. Until they do, they must accept their culpability in creating an overheated planet with millions of climate refugees."

Only at the "illegal" plant at Chernobyl, which was designed to make plutonium for bombs, with electricity being a by-product, has anyone died from radiation from nuclear power, but we've had tens of millions of coal, gas and petroleum-related deaths. Furthermore, our reactors, by generating electricity from the 20,000 Russian warheads we purchased in the *Megatons to Megawatts* program, have become the ultimate in weapons-reduction techniques.

www.scientificamerican.com/article.cfm?id=the-human-cost-of-energy
http://tinyurl.com/kn22qcn
http://tinyurl.com/m5qp8vf
http://cen.acs.org/articles/91/web/2013/04/Nuclear-Power-Prevents-Deaths-Causes.html
www.forbes.com/sites/jamesconca/2012/06/10/energys-deathprint-a-price-always-paid/

So, what about Three Mile Island, Chernobyl and Fukushima? We'll examine each of them, but it is important to realize that <u>nuclear plants have been supplying 15% of the world's electricity, *while creating no CO2,* for 16,000 reactor-years of almost accident-free operation</u>. We should also recall that the reactors that have powered our nuclear navy for more than 50 years have similar safety records.

Three Mile Island

In March, 1979, two weeks after the release of the popular movie, *The China Syndrome*, a partial meltdown of a reactor core due to a "stuck" coolant valve and design flaws that confused the operators, caused mildly radioactive gases to accumulate inside one of the reactor buildings.

After the gases were treated with charcoal, they were vented, and a small amount of contaminated water was released into the Susquehanna River. No one died or was harmed.

However, when an AP reporter described a "bubble" of hydrogen inside the reactor building in a way that led people to think that the plant was a "hydrogen *bomb*," many residents fled, which caused even more harm than the accident.

In fact, radiation exposure from Three Mile Island was far less than the amount of radiation that pilots and airline passengers receives during a round-trip flight between New York and Los Angeles. Furthermore, in the following decades, more than a dozen studies have found *no short or long-term ill effects for anyone,* whether they were downwind or downstream from the plant or at it – and since then operator training and safety measures have greatly improved.

Despite all of the fear and panic, nothing happened. No one died, and no one got cancer, but the media-hyped event at Three Mile Island came very close to shutting down all progress in American nuclear power. Because of the radiophobia generated by our sensation-seeking press and fervent greens, neither of whom bothered to check the facts, many proposed reactors were replaced by coal plants, and in the following decades, pollution from those coal plants killed at least 100,000 Americans.

CHERNOBYL

In 1986, during a test ordered by Moscow that involved disabling the safety systems, a portion of the core of the reactor, which had design hazards *not present* in Western reactors, was inadvertently exposed. As Spencer Weart wrote in The Rise of Nuclear Fear, "In short, for Soviet reactor designers, safety was less important than building 'civilian' reactors that could produce military plutonium if desired, and building them cheaply."

This negligence led to a steam/hydrogen explosion that released radioactive gases into the atmosphere because the reactor had a flammable core and **no containment structure**. In contrast, every U.S. reactor contains no flammables. Each has a reinforced concrete containment structure that can survive an airliner hit, and every plant is strictly regulated by the NRC. There has never been a source of energy as safe or kind to the environment as nuclear power, and the reason for the safety is regulation.

Every responsible nation similarly regulates its nuclear power plants and shares information and training practices via international agencies. This cooperation, which was expanded after Three Mile Island, resulted in so many improvements that civilian nuclear power climbed from 60% uptime in the sixties to at least 90% today.

Chernobyl, which was primarily a facility for making plutonium for nuclear weapons, but also produced electricity for Pripyat, was long judged to be dangerous by scientists outside of the Soviet Union. (For three days, the Russian authorities hid the disaster and delayed evacuating the area, coming clean only when radiation readings across Europe began to rise. (In addition, the government failed to distribute iodine tablets, which could have protected thousands from airborne Iodine-131, which is readily absorbed by the thyroid, particularly in the young. (A body with an abundance of benign I-127 is less likely to absorb I-131.)

Chernobyl, which failed due to bad design, poor training and a system that forbade operators from sharing essential information about reactor problems, is the *only* "civilian" reactor accident where radiation directly killed anyone. Initially, approximately eighteen firefighters died from intense radiation. Yet, with design changes and proper procedures, several similar reactors still operate in the former Soviet Union.

*According to a study by 100 scientists from eight United Nations agencies, "Chernobyl produced additional 50 deaths over the following **twenty** years."* Most died soon after the accident. However, that's a tiny fraction of the deaths caused by burning coal or oil or natural gas.

Furthermore, the deformed and brain-damaged "Chernobyl children" that sensation-seeking TV programs feature are no different from similarly afflicted children elsewhere in Russia who received *no fallout*, but that information is never provided by many anti-nuclear activists and the media. (Since Chernobyl, cancer rates in the Ukraine have been about 2/3 of the rate in Australia.)

Finally, because of the erroneous (and dangerous) LNT theory and many dire predictions from influential people like Helen Caldicott (See Chapter 11.), thousands of badly frightened European women endured voluntary abortions because they had become convinced that they were carrying monster babies.

http://dailym.ai/2mLRQPV **Chernobyl today**

Fukushima

Tepco's Fukushima Daiichi reactors began operation in 1971 and ran safely for 40 years, generating huge amounts of electricity without creating any CO_2 or air pollution, but then, in 2011, came a *record-setting* earthquake.

During the earthquake, which shifted Honshu, Japan's main island, 8 feet eastward, all of Japan's 52 reactors began to shut down properly, including those at Fukushima.

However, the quake destroyed the plant's connections to the electrical grid, which required emergency generators to power the systems that cooled the reactors.

Although three of Tepco's six nuclear reactors were not operating when the quake struck, five were eventually doomed because:

1. In 1967, Tepco removed 25 meters from the site's *35-meter seawall* to ease bringing equipment ashore.

2. Tepco replaced the original seawall with only a **six-meter** *seawall*.

3. The Japanese government advised Tepco to raise it, but Tepco declined – and the government did nothing.

4. Tepco had inexplicably placed five of its six emergency generators in the *basements*.

5. The tsunami flooded all but # 6.

6. Batteries powered the controls for about 8 hours, and then failed. Without coolant, meltdown was assured.

Reactors 1 - 4 are useless, and number 5 is damaged, but reactor 6 was unaffected because its back-up equipment was intelligently sited well above the tsunami's reach. Reactor 6 is capable of producing power, but it has not been started, largely because of the anti-nuclear hysteria fanned by most of the Japanese press.

http://www.whoi.edu/page.do?pid=127297

There were warnings: All along the coast, ancient "Sendai stones" had been warning residents to avoid building below 150 feet above sea level for centuries.

The Onagawa nuclear plant, which was *closer* to the epicenter of the quake, also survived the quake, and its *45-foot* high seawall easily blocked the tsunami. The tsunami took more than 15,000 lives, but the Fukushima failure directly took the lives of just two workers who drowned.

Japan responded by closing its nuclear plants – a foolish move that has required the country to spend $40 billion per year on liquefied natural gas plus billions more for coal, which has created huge amounts of greenhouse gases. Another $11 billion per year has been spent to maintain their perfectly functional-but-idle reactors.

Nuclear power has been tarred by the Fukushima Daichi disaster, but the failure was NOT the fault of nuclear power. It was caused by repeated corporate lying, record falsifying and penny-pinching, by the lack of government enforcement of seawall height, by building too low to the ocean, and by installing backup generators in easily flooded basements.

Blaming nuclear power for Fukushima is like blaming the train when an engineer derails it by taking a turn at 70 mph that is posted for 30.

In 2015, the usually reliable Amy Goodman reported that a class action suit had been filed by several sailors who had served on the USS REAGAN. In her article, she described their symptoms, which they blamed on being exposed to radiation, but she failed to provide any depth.

https://www.youtube.com/watch?v=Zw33AVqzQxA)

A few days later, Goodman's article was read by **Captain Reid Tanaka,** a United States navy professional with considerable expertise in nuclear matters who had been intimately involved during the meltdown – and Captain Tanaka presented a very different view:

"I was in Japan, in the Navy, when the tsunami struck and because of my nuclear training, I was called to assist in the reactor accident response and served as a key advisor to the US military forces commander and the US Ambassador to Japan. I spent a year in Tokyo with the US NRC-led team to assist TEPCO and the Japanese Government in battling through the casualty.

"My command (CTF 70) was the direct reporting command for the REAGAN (where we had control over REAGAN'S assignments and missions) and were in direct decision-making with REAGAN'S Commanding Officer and team. I don't qualify to be called an "expert" in reactor accidents..., but I am well informed enough to know where my limits are and to see through much of the distortions on this issue....

"A Google search will tend to drive people to alarmist websites and non-technical news reports, but you could also find the dull, technical (yet truthful) places such as the IAEA or DOE...

"Numerous bodies of experts have weighed in and provided assessments and reports. A couple are quite critical of TEPCO and the Japanese nuclear industry and regulators.

"... the biggest problem the public has is ... being able to distinguish the science-based, objective reports from the alarmist and emotionally charged positions that get the attention of the press, some of whom are self-proclaimed experts in some fields but NOT nuclear power: Dr. David Suzuki and Dr. Michio Kaku. Neither understand spent fuel, nor the condition of spent fuel pools....

"Dr. Suzuki is an award-winning scientist and a champion for the environment, but he is lacking any real understanding of spent fuel or radioactivity. "Bye-bye Japan?" A headline grabbing sound-bite, but the math just doesn't work...

"[Sometimes] the true experts cannot give a simple answer because there isn't one, while those who have no science to back their claims have no compunction in saying the sky is falling and everyone else is lying.

"For the Navy, the contamination caused by Fukushima created a huge amount of extra work and costs for decontaminating the ships and our aircraft to "zero", but [there was] no risk to the health of our people.

"REAGAN was about 100 miles from Fukushima when the radiation alarms first alerted us to the Fukushima accident. Navy nuclear ships have low-level radiation alarms to alert us of a potential problem with our onboard reactors. So, when the airborne alarms were received, we were quite surprised and concerned. The levels of airborne contamination were small, but caused a great deal of additional evaluation and work. REAGAN's movements were planned and made to avoid additional fallout. Sailors who believe they were within five miles or so, were misinformed. Japanese ships were close; the REAGAN was not....

"There are former SAILORS who are engaged in a class-action suit against TEPCO for radiation sickness they are suffering for the exposure they received from Operation Tomodachi. The lead plaintiffs were originally sailors from REAGAN but now have expanded to a few other sailors from other ships. Looking at the claims, I have no doubt some of the SAILORS have some ailments, but without any real supporting information (I haven't seen ANY credible information to that end), I do not believe any of their ailments can be directly attributable to radiation—fear and stress related, perhaps, but not radiation directly. Radiation sickness occurs within a "minutes/hours" timeframe of exposure and cancer occurs in a "years" timeframe. These SAILORS were not sick in either of these windows. I believe that many of them believe it, but I also believe most are being misled."

The closure of Japan's nuclear plants and its increased use of imported liquefied natural gas put an end to Japan's long-standing trade surplus. But in 2015, bowing to financial realities and because of diminishing fear, Japan restarted the second of its reactors, with plans to add more.

Shortly thereafter, the U. S. media and many of the "Green" organizations began to report that a Fukushima worker had been "awarded compensation and official acknowledgment that his cancer [leukemia] was caused by working in the reactor disaster zone." That's wrong, and competent journalists who do adequate research should know it. Here are the facts:

The worker received a workman's comp benefit package because he satisfied the statutory criteria stipulated in the 1976 Industrial Accident Compensation Insurance Act, which says that workers who are injured or become ill while working or while commuting to and from work, can receive financial aid and medical coverage. The worker spent 14 months at F. Daiichi. (October, 2012 to December 2013.)

In late December 2013, the worker felt too ill to work, so he went to a doctor, and was diagnosed with acute leukemia in January, 2014. No link was made between his occupational exposure and his cancer. <u>In addition, because the latency period between radiation exposure and the onset of leukemia is 5 to 7 years, the worker did not get cancer from working at Fukushima. It was, in fact, a pre-existing condition</u> that was exploited by opponents of nuclear power who routinely repeat convenient-but-wrong stories because being honest and accurate takes time, knowledge and integrity.

In early 2016, anti-nuclear zealots began to fear-monger about the effects of Cesium-134 on fish while ignoring reports from NOAA and the Japanese government that stated, "Radioactive Cesium in fish caught near F. Daiichi continues to dwindle.

"Of the more than 70 specimens taken in October, only five showed any Cesium isotope 134, the 'fingerprint' for Fukushima Daiichi contamination… The highest Cs-134 concentration was [associated] with a Banded Dogfish, at 8.3 Becquerels per kilogram. Half of the sampled fish had detectible levels of Cs-137, but <u>all were well below Japan's limit of 100 Bq/kg</u>…."

These amounts are tiny, and the particles emitted from the Potassium-40, which we all contain, are more potent than the Cesium-137 emissions that many greens apparently fear.

Regarding the risk from the remaining reactor material that many greens agonize over, **Dr. Alex Cannara** subsequently wrote,

"As of late **2013,** the spent fuel at Fukushima was 30 months old. That means that the rods and the fuel pellets within them are able to be stored in air. If any rods had never been in a reactor core, they have no fission products in them and are perfectly safe to take apart by hand.

"So, what do we have at Fukushima? We have some melted core materials (corium), which can be entombed. We have water containing a small amount of fission products like Cesium. And, we have a bunch of fuel assemblies that are very radioactive because of their internal creation of fission products when they were in their reactor cores. (No fission products are created when rods are out of cores, or in pools or dry air storage.)

"Since the rods are at least 30 months out of fission-product production, one can see how quickly they've lost the need for cooling, and the reduction in their radioactivity. Nuclear power has for its entire life, been the safest form of power generation. The EPA estimates that we lose more than 12,000 Americans every year to coal emissions. The Chinese lose 700,000, and the Indians - 100,000. To delay building nuclear power plants will cause diseases and deaths that could easily be avoided."

www.energyfromthorium.com/javaws/SpentFuelExplorer.jnlp

Andrew Daniels

"A nuclear power plant that melts down is less dangerous than a fossil fuel plant that is working correctly. [Because of their toxic ashes and emissions.] Fukushima illustrates that even a meltdown that penetrates containment is very little danger to the public when a few basic precautions are taken."

http://nuclearprogress.org/how-fukushima-made-me-a-nukie/

Chapter 7
What's the Fossil Fuel Record Safety and Death-prints
The Fossil Fuel Record

millions of air pollution deaths

Because the carbon industries have been very heavily subsidized, one might expect them to have exemplary safety and social records, but one would be wrong!

According to **the Guardian** (10-17-16)

"Fossil fuel companies are benefitting from global subsidies of $5.3 trillion a year, equivalent to $10 million a minute every day, according to a startling new estimate by the International Monetary Fund. The IMF noted that existing fossil fuel subsidies overwhelmingly go to the rich, with the wealthiest 20% of people getting six times as much as the poorest 20% in low and middle-income countries...."

https://www.theguardian.com/environment/2015/may/18/fossil-fuel-companies-getting-10m-a-minute-in-subsidies-says-imf

In 2006, the Sago coal mine disaster killed 12. A few years later, a West Virginia coal mine explosion killed 29. In May 2014, 240 miners died in a Turkish coal mine.

The ash derived from burning coal averages *80,000 pounds* per American lifetime. <u>Compare that amount to two pounds of nuclear "waste" for the same amount of power.</u> The world's 1,200 largest coal-fired plants cause 12,000 premature American deaths every year plus hundreds of thousands of cases of lung and heart diseases in the U. S.

Generating the 20% of U.S. electricity with nuclear power plants saves our atmosphere from being polluted with 177 million tons of greenhouses gases *every* year, but despite the increasing consequences of Climate Change and Ocean Acidification, the burning of fossil fuels (carbon) for power is still rising.

"Coal-fired plants expel mercury, arsenic, uranium, radon, cyanide and harmful particulates while exposing us to 100 times more radiation than nuclear plants that create no CO_2. In fact, coal ash is more radioactive than any emission from any operating nuclear plant." (**Scientific American**, 12-13-07.) <u>https://www.southernenvironment.org/news-and-press/news-feed/duke-energy-pleads-guilty-to-environmental-crimes-in-north-carolina</u>

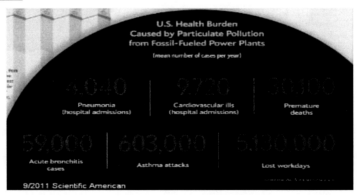

Every year, we store 140 million tons of coal ash in unlined or poorly lined landfills and tailing ponds. In 2008, five million tons of toxic ash burst through a Tennessee berm (see below), destroying homes and fouling lakes and rivers. Coal-fired power plants leak more toxic pollution into America's waters than any other industry. (A June 2013 test found that arsenic levels leaking from unlined coal ash ponds were 300 times the safety level for drinking water.)

And in 2014, North Carolina's Duke Energy's plant (now bankrupt) "spilled" 9,000 **tons** of toxic coal ash sludge into the Dan River, but as of January, 2017, the cleanup" was still proceeding. Why do they always "spilled" – never "gushed" or "erupted."

Coal companies like to promote their supposedly "clean coal," which really means "not quite so filthy," but despite making an attempt at carbon capture and storage (CCS) at a new power plant in Saskatchewan, the plant has been deemed a failure:

Technology to Make Clean Energy from Coal is Stumbling in Practice

NYT by IAN AUSTEN   3-29-2016 - OTTAWA

"An electrical plant in Saskatchewan was the great hope for industries that burn coal.

"In the first large-scale project of its kind, the plant was equipped with a technology that promised to pluck carbon out of the utility's exhaust and bury it, transforming coal into a cleaner power source. In the months after opening, the utility and the provincial government declared the project an unqualified success, but the $1.1 billion project is now looking like a dream.

"Known as SaskPower's Boundary Dam 3, the project has been plagued by shutdowns, has fallen way short of its emissions targets, and faces an unresolved problem with its core technology. The costs, too, have soared, requiring tens of millions of dollars in new equipment and repairs.

"At the outset, its economics were dubious," said Cathy Sproule, a member of the legislature who released confidential internal documents about the project. "Now they're a disaster...."

Please note: The word "efficiency" in this book means the amount of electricity actually produced over an extended period by the various means of generating power compared to their maximum power rating, which is the figure often used to sell the project. For conventional nuclear reactors, this figure is 90%, but for windmills and solar, it is 30% at best. (It is not intended to describe *thermal* efficiency, which is a separate matter.)

Even modern, 75% efficient coal-burners with thirty-year lifespans can't compete with nuclear plants that have lifespans of 50 or more years, and provide CO2-free power at 90% efficiency. And the new plants are even safer, with some occasionally being 98% efficient. In addition, our seemingly endless coal reserves will last just 100 years at best, and we've steadily been increasing consumption. As we "decarbonize", we will require increasing amounts of electricity, and the only source of economical 24/7 power will be safe, super-efficient, CO2-free nuclear reactors, including modular MSRs.

When a gas pipeline exploded in San Bruno, California, eight people died, 35 homes were leveled and dozens more were damaged. (In 2016, a federal government report stated that natural gas explosions cause heavy property damage, often with deaths, about 180 times per year – that's every other day.")

www.scientificamerican.com/article.cfm?id=the-human-cost-of-energy

In 2010, British Petroleum's Deepwater Horizon disaster in the Gulf of Mexico "spilled" 200 million gallons of oil and killed eleven workers. Prior to that, an explosion at a Texas BP refinery killed fifteen. And B P, which was also involved in the Exxon Valdez "spill" in Alaska's Prince William Sound, is just one of the many oil companies that we subsidize with $2.4 billion every year.

http://www.newscientist.com/article/mg20928053.600-fossil-fuels-are-far-deadlier-than-nuclear-power.html#.VK4ftS7CaSq

Also in 2010, an Enbridge pipeline ruptured in Michigan, eventually "spilling" more than a million gallons of tar sands crude oil into the Kalamazoo River. (When monitors at Enbridge's Albert headquarters reported that the pressure in the line had fallen to zero, control room staff dismissed the warning as a false alarm, and instead of shutting the pipeline down, cranked up the pressure in the line twice, which worsened the disaster. In 2016, Enbridge's "cleanup" was still incomplete.

In 2013, a spectacular train wreck dumped 2 million gallons of North Dakota crude into Lac Megantic, Quebec, killing

47 residents and incinerating the center of the town – but that's just another page in the endless petroleum tale that includes Exxon's disastrous, 2016 "spill" in Mayflower, Arkansas, that inexplicably received scant notice from the press.

And in November, 2013, a train loaded with 2.7 million gallons of crude oil went incendiary in Alabama, followed in December by a North Dakota conflagration.

2014 began with a fiery derailment in New Brunswick, Canada, and in October 2014, 625,000 liters of oil and toxic mine-water were "spilled" in Alberta. Jm

July, August and September brought Alberta's autumn, 2014 total to 90 pipeline "spills." 2015 brought four, fiery oil train wrecks just by March, and 2016 delivered two Alabama pipeline explosions, one close to Birmingham.

In late 2015, California's horrific, Aliso Canyon methane "leak" (think "geyser") erupted, spewing 100,000 tons of natural gas, the equivalent of approximately 3 billion gallons of gasoline or adding 500,000 cars to our roads for a year. The Southern California Gas Company finally managed to throttle the geyser in February, 2016. Incidentally, Aliso's 100,000 tons of leakage is just 25% of California's "allowed" leakage, which is an indication of the political power of the natural gas industry. (Five months later, a new headline appeared: *"Massive Fracking Explosion in New Mexico")*

The Aliso "leak" caused the loss of 70 billion cubic feet (BCF) of gas that California utilities count on to create electricity for the hot summer months. As a consequence, the California Independent Service Operator, which manages California's grid, estimated that due to Aliso, 21 million customers should expect to be without power for 14 days during the summer.

According to **Business Insider** (July, 2016), "SoCalGas uses Aliso Canyon to provide gas to power generators that cannot be met with pipeline flows alone on about 10 days per month during the summer."

However, during the summer, SoCalGas also strives to fill Aliso Canyon to prepare for the winter heating season. State regulators, however, subsequently ordered the company to reduce the amount of gas in Aliso to just 15 BCF and use that fuel to reduce the risk of power interruptions in the hot summer months of 2016. Fortunately, State regulators have also said that they won't allow SoCalGas to inject fuel into the facility until the company has inspected all of its 114 storage facilities.

The Aliso canyon disaster wiped out all of the state's debatable greenhouse gas reductions from its wind and solar systems. And in July, 2016, California officials reported leakage at a San Joachim County storage facility that was "similar to, or slightly above, background levels at other natural gas storage facilities." In other words, ALL of these sites are leaking!

Dr. Alex Cannara, a California resident writes, "Combustion sources [unlike nuclear power], aren't burdened with their true costs. Gas, for example, is not cheaper than nuclear or anything else. In 2016, our allowed leakage wipes wind/solar out by 4 times. In other words, 'renewables' in a gas state like California wipe out their benefits every 3 months because they depend on gas for most of their nameplate ratings. The Aliso storage was largely used to compensate for 'renewables' inevitable shortfall.

"The most important combustion cost is the *unlimited* downside risk of its emissions for the entire planet, but in February 2016, our CEC approved 600MW of *added* gas burning in the San Diego region simply because San Onofre wasn't running, due to possibly corrupt actions by SoCla Gas, SCE, Sempra Energy and Edison Intl.

"Such practices were prevented for 75 years by the 1935 PUHCA, but the Bush-Cheney administration signed its repeal in 2005 after decades of carbon combustion-interest lobbying. However, a few states – not California – passed legislation to correct for the 2005 PUHCA repeal."

There's more: In August, 2016, the **Pennsylvania Department of Environmental Protection** admitted that oil and gas production in the state emitted as much methane as Aliso Canyon. The Aliso "leak" was deemed a disaster, but the hundreds of equally damaging Pennsylvania "leaks" were considered business as usual.

Finally, also in August, 2016, a thirty-inch pipeline exploded in southeast New Mexico, killing five adults and five children while leaving two other adults in critical condition in a Lubbock, Texas hospital. All of this could have been avoided if, instead of pursuing intermittent, short-lived, <u>carbon-dependent</u> windmills and solar panels (Chapters 8 and 9), we had expanded safe, CO2-free nuclear power.

Dr. Wade Allison, in *Nuclear is For Life,* wrote: "Critics of *civilian* nuclear power use what they fear *might* happen due to a nuclear failure – but never has – but ignore other accidents that have been far worse:

"The 1975 dam failure in China that killed 170,000.

The 1984 chemical plant disaster in Bhopal, India where 3,899 died and 558,000 were injured;

The 1889, Johnstown. PA flood that drowned 2,200;

The 1917 explosion of a cargo ship in Halifax, N. S. where 2,000 died and 9,000 were injured

The 1917 cargo ship explosion in Halifax, N. S., where 2,000 died and 9,000 were injured;

The 2014 coal mine accident in Turkey that took 300 lives;

The 2015 warehouse explosion at Tianjin, China that cost 173 lives.

"The list seems endless, but no one advocates destroying dams, stopping mining and closing chemical plants. The way the world reacted to the Fukushima accident has been a disaster, with many huge consequences to the environment, but the accident itself was not."

Mike Conley

"In California, defective, Japanese-built steam generators at the San Onofre nuclear plant could have been replaced for about $600 million, but the plant is being decommissioned at a cost of $4.5 billion because of Fukushima and anti-nuclear zealotry. Alternatively, the entire plant could be replaced with two, modern, CO_2-free AP-1000 reactors for about $14 Billion."

In 2016, a judge reopened a case related to the shutdown of the San Onofre nuclear power plant because Sempra Energy and Edison International may have exploited the Bush/Cheney repeal of the 1935 Public Utilities Holding Company Act, which had protected utility customers for 70 years.

I repeat, NO ONE has died from radiation created by *civilian* nuclear power, but more than 2.000,000 die prematurely every year from the burning of coal, gas, wood and oil.

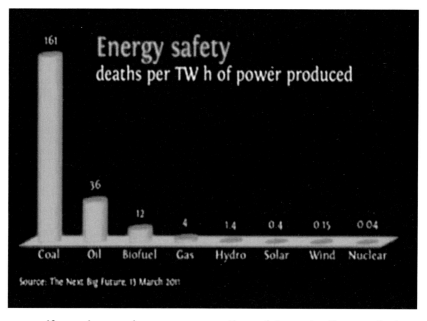

If we give nuclear power a rating of 1 on deaths per kwh, coal is 4,000 times worse, oil comes in at 900 and natural gas gets a 100. I repeat, NO ONE has died from radiation created by commercial nuclear power production in Western Europe, Asia or the Southern and Western hemispheres, but more than two million die every year from the burning of coal, gas and oil.

Chapter 8

Powering Ships and Desalination

What's a Light Water Reactor?

"Waste" Management.

What's a MSR? What's a LFTR?

Cargo ships emit more air pollution than all of the world's cars, but we don't power container ships with CO2-free nuclear power because we are worried about nuclear proliferation. However, if we would equip these ships with proliferation-*resistant* reactors, we could save seven million barrels of oil per day, eliminate 4% of our greenhouse gas emissions and replace those huge fuel tanks with profitable cargo.

Propelling one of our immense aircraft carriers at 27 mph for 24 hours requires only **three pounds** of nuclear fuel – the equivalent of 400,000 gallons of diesel fuel. (Burning just 100 gallons of diesel creates one ton of carbon dioxide, and just 16 supertankers or large cargo ships create as much pollution as all the cars in the world.)

California's drought-stricken Central Valley, which was a dry savanna before "civilization" arrived, is more than 10 trillion gallons per year behind in precipitation, and San Diego County alone needs at least 700 million gallons per year. Fortunately, there is a remedy, but that remedy will require an abundance of carbon-free electricity created by nuclear power.

http://earthsky.org/earth/tree-ring-study-shows-californias-drought-worst-in-1200-years

In May, 2016, Pacific Gas and Electric (PG&E) announced that the Diablo Canyon nuclear power plant will begin producing 1.5 million gallons per day of desalinated water. There should be many more plants like Diablo, and there would be, but for the opposition of anti-nuclear zealots, whose efforts have helped raise CO_2 levels and worsened Climate Change. If the greens succeed in closing California's San Onofre nuclear power plant, California will need to burn huge volumes natural gas (methane) to replace San Onofre's 2.4 billion watts of *carbon-free* electricity.

The non-nuclear Carlsbad desalination plant currently produces about 50 million gallons of fresh water per day, providing just 7% of San Diego's needs with 40 MW of power, but to serve all of California, we would need 140 Carlsbads. In comparison, just three Diablo Canyon plants could be equally productive – and with a much smaller "carbon footprint", which is a measure of the amount of carbon burned per project from shovel-in-the-ground to power-at-the-outlet.

Why do we persist with carbon fuels when six uranium pellets the size of your little finger, which create no CO2, contain as much energy as 3 tons of coal or 60,000 cubic feet of natural gas? Just a fistful of Uranium can run all of New York City for an hour, and the waste products are far less than that. The Excel Energy plant at Becker, Minnesota, turns <u>60 million pounds of coal per day into CO2,</u> but less than <u>100 pounds</u> of uranium would produce the same amount of electricity without creating any CO2.

How does a water-cooled, uranium-powered Light Water Reactor (LWR) work?
What are its pluses and minuses?

Some people claim that uranium mining is especially dangerous because the ore is radioactive, but they are mistaken. The radiation level just one foot from a drum of uranium oxide is only about 20% of the cosmic radiation level that passengers experience on a jet flight – and the ore from which the oxide was derived is even less hazardous.

In a conventional reactor, uranium pellets containing 3.5% to 5% U-235 are sealed in 25,000 or more 12-foot-long zirconium tubes. Within those tubes, the U-235 emits neutrons that sustain a chain reaction that releases huge amounts of heat. Because the water surrounding the tubes is heated to 600 degrees F, it is "kept" at 2,700 psi to prevent it from boiling.

This super-heated water is circulated through a heat-exchanger to make steam in a separate plumbing loop that powers a turbine, which spins a generator. And because the super-heated, pressurized water would explosively expand 1,000 times if there were a leak, a huge, immensely strong, concrete containment dome encloses the reactor so that steam and other gases cannot escape. Once started, a LWR can run constantly for about five years without refueling.

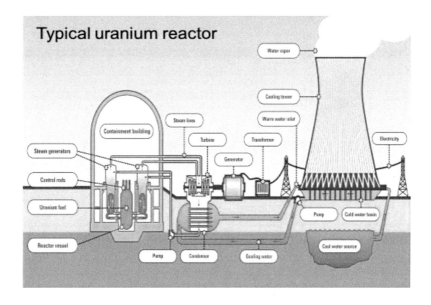

What about the waste?

Nuclear power plants are required to contain every speck of their waste. If you were to get all the electricity for your lifetime from conventional uranium reactors, your share of the waste would weigh just **two** pounds – but only a small portion of that would be hazardous long term.

During fission, reaction waste products called actinides accumulate in the pellets, which become cracked. Other reactions that involve the zirconium rods and the super-heated water can produce hydrogen, which, if not eliminated, can explode, as it did at Fukushima.

As the pellets accumulate actinides and other by-products, they become inefficient, and must be replaced during a multi-day shut-down during which the assemblies are moved by remotely operated cranes to storage pools filled with water. There, the water absorbs neutrons and other radiation to keep the slowly decaying fuel from overheating. After a few years, the pellets' radioactivity has decreased enough to allow them to be safely moved to air-cooled, dry cask storage on the plant site, which is our current, long term solution.

Spent fuel bundles stored under water at a nuclear plant.

About 95% of the spent fuel created by a conventional plant is ordinary Uranium 238 and 235 that can serve as fuel if the actinides are removed. This "waste" can be worth billions as *modern* reactor fuel, and it also contains isotopes that are essential for nuclear medicine and for powering satellites.

Fortunately, a 2014 study showed that (after removing the U-238 and 235), which are useful, mixing plutonium-contaminated waste with blast furnace slag to convert it to "glass" reduces its volume by 90% and locks the plutonium into a stable end product. Heavily nuclear France, however, uses a recycling program that greatly reduces its volume and the length of time it must be stored. As a consequence, all of France's multi-decade waste could be stored on just one basketball court.
https://www.youtube.com/watch?v=IzbI0UPwQHg
http://bigthink.com/philip-perry/scientists-turn-nuclear-waste-in-diamond-batteries-thatll-last-for-thousands-of-years

In comparison, all of the nuclear waste generated in the U.S. since the fifties could be stored on one football field in nine-foot tall, self-ventilating, concrete containers. After just 40 years of storage, only about one thousandth as much radioactivity remains as when the reactor was turned off for fuel replacement. Therefore, only a small portion requires long term storage or recycling.

However, because recycling can retrieve plutonium isotopes from the waste, some of which are used for making bombs, President Carter closed our only recycling plant during the Cold War in an attempt to placate Russian fears that we'd be making nuclear bombs.

Unfortunately, there is another reason: the anti-nuclear crowd has promoted radiophobia so effectively that many voters and legislators have refused to build even the new, super-safe, highly efficient reactors that can consume our stored waste as fuel. www.energyfromthorium.com/javaws/SpentFuelExplorer.jnlp
http://www.foxbusiness.com/markets/2016/11/15/general-electric-and-southern-company-team-up-to-power-planet-with-nuclear.html

What's a MSR?

Molten Salt Reactors avoid many of the disadvantages of conventional reactors.

Fluoride salts also don't break down under high temperatures or high radiation, and they lock up radioactive materials, which prevents them from being released to the environment. As noted earlier, Alvin Weinberg's Oak Ridge MSR ran successfully for 22,000 hours during the sixties. However, the program was shelved, partly because we preferred Admiral Hyram Rickover's water-cooled reactors. In a Molten Salt Reactor, the uranium (probably thorium in the future), is dissolved in a *liquid* fluoride salt. Note: Although fluorine gas is corrosive, fluoride salts are not.

Schematic of a Molten Salt Reactor

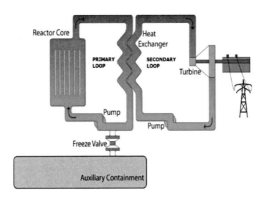

When uranium or thorium is combined with a liquid fluoride salt, there are no pellets, no zirconium tubes, no water and, therefore, no hydrogen that might explode.

The fluid that contains the nuclear fuel is also the heat-transfer agent, so no water is required for cooling. MSRs are also more efficient than conventional plants because the temperature of the molten salt is about 700 degrees C (1300 F), whereas the temperature of the water in a conventional reactor is about 330 degrees C, (620 F), and higher heat creates more high pressure steam to spin the turbines.)

That extra heat can be used to create more electricity, desalinate seawater, split water for hydrogen fuel cells, make ammonia for fertilizer and even extract carbon from the air and our oceans to make gasoline and diesel fuel. *In addition, MSRs can consume 96% of our 68,000 tons of stored uranium "waste" and the fissile material in our thousands of nuclear bombs.*

http://www.huffingtonpost.com/sherry-ray/post_12583_b_10942568.html

Because the boiling point of the molten salt is 1400 degrees C (2550 F), and most of the MSR designs do not need to be water-cooled, those versions don't risk a steam explosion that could propel radioactive isotopes into the environment. And because MSRs operate at atmospheric pressure, no huge, re-enforced concrete containment dome is needed.

When the temperature of the liquid salt fuel in the core rises as the chain reaction increases, the molten fuel expands, which decreases the density of the fuel and slows the rate of fission, which prevents a "runaway" reaction.

As a consequence, a MSR is inherently stable or "self-governing," and because the fuel is a liquid, it can easily drain *by gravity* into a large containment reservoir. In fact, the results of a fuel "spill" from a MSR would be measured in square yards, not miles.

In the event of an electric power outage, a refrigerated salt plug at the bottom of the reactor automatically melts, allowing the fuel to drain into a tank, where it spreads out, cools and solidifies, stopping the reaction. In effect, MSRs are <u>walk-away safe.</u> Even if you abandon a MSR, the fuel will automatically drain, cool down and solidify without human intervention. <u>If the Fukushima reactor had been a MSR, there would have been no meltdown, and because radioactive by-products like cesium, iodine and strontium bind tightly to stable salts, they would not have been released into the environment.</u>

USEFUL MSR BYPRODUCTS

Besides producing CO_2-free power, the fissioning U-233 in a Molten Salt Reactor creates essential industrial elements that include xenon, which is used in lasers and even in anaesthesia, neodymium for high-strength magnets, strontium, medical molybdenum-99, zirconium, rhodium, ruthenium, palladium, iodine-131 for the treatment of thyroid cancers and bismuth-213, which is used for targeted cancer treatments.

What's a LFTR?

A thorium-fueled MSR is a
Liquid Fluoride Thorium Reactor -
a **LFTR** – pronounced **LIFTER**

A Lifetime of power in the palm of your hand

With a half-life of 14 billion years, Th-232 is one of the safest, least radioactive elements in the world. Thorium-232 emits harmless alpha particles that cannot even penetrate skin, but when it becomes Th-233 in a Molten Salt Reactor, it's a very potent source of power. Sunlight, living at high altitude and the emissions from your granite countertop or a coal-burning plant are more hazardous than thorium-232.

LFTRs are even more fuel-efficient than uranium-fueled MSRs, and they create little waste because a LFTR "burns" 99% of the thorium-232. Conventional reactors (LWRs), consume just 3% of their uranium. That's like burning a tiny part of a log while the rest gets contaminated with chemicals you must store for years.

Just <u>one pound of thorium creates as much electricity as 1700 tons coal, so replacing coal-burning plants with LFTRs would eliminate one of the largest sources of climate change</u>. That same pound (just a golf ball-size lump), can yield all the energy an individual will ever need, and just one cubic yard of thorium can power a small city for at least a year. In fact, if we were to replace ALL of our carbon-fueled, electrical power production with LFTRs, we would eliminate 35 to 40% of all man-made greenhouse gas production.

From 1977 to 1982, the LWR at Shippingport, PA was powered with a Thorium/Uranium mix, and when it was eventually shuttered, the reactor core was found to contain about 1% <u>more</u> fissile material (U233/235) than when it was loaded. (Thorium has also fueled the Indian Point 1 facility and a German reactor.)

India, which has an abundance of thorium, is planning to build Thorium-powered reactors, as is China, while we struggle to overcome our public's unwarranted fear of nuclear power. And in April, 2015, a European commission announced a project with 11 partners from science and industry to prove the innovative safety concepts of the Thorium-fueled MSR and deliver a breakthrough in waste management.

Please see ***Thorium: the last great opportunity of the industrial age*** - by David Archibald

http://wattsupwiththat.com/2015/05/16/thorium-the-last-great-opportunity-of-the-industrial-age/

https://www.nytimes.com/2016/12/21/opinion/to-slow-global-warming-we-need-nuclear-power.html?_r=1

Supplies

Thorium ore (Th-232) is four times as plentiful as uranium ore, and Th-232, besides being entirely useable, is 500 times more abundant than uranium's fissile U-235 isotope. Even at current consumption rates, uranium fuels can last for a few centuries, but thorium could power our world for many thousands of years.

Just 1,000 tons of thorium is equivalent to 47 <u>billion</u> barrels of oil or 460 <u>billion</u> cubic meters of natural gas. We have about 400,000 tons of thorium ore, and we don't even need to mine thorium because our *Rare-Earth Elements* plant receives enough thorium to power our entire country every year. Australia and India tie for the largest at about 500,000 tons, and China is well supplied.

Waste and storage.

Due to their fuel efficiency, LFTRs create only 1% of the waste that conventional reactors produce, and because only a small part of that waste needs storing for just 400 years – not the thousands of years that LWR waste requires - repositories much smaller than Yucca mountain would easily suffice.

Furthermore, LFTRs can run almost forever because they produce enough neutrons to make their own fuel, and the radiotoxicity from LFTR waste is 1/1000 that of conventional reactor waste. That said, the best way to eliminate most nuclear waste is to stop creating it with conventional reactors and replace them with modern reactors like MSRs or LFTRs that can consume stored waste as fuel.

With no need for huge containment buildings, MSRs can be smaller, in both size and power, than current reactors, so ships, factories, and cities could have their own power source, thus creating a more reliable, efficient power grid by cutting long transmission line losses that can run from 8 to 15%.

Unfortunately, few elected officials willl challenge carbon industries that provide millions of jobs and wield great political power. As a consequence, thorium projects have received little help from the government, although China and Canada are moving toward thorium, <u>and India already has a reactor that runs on 20% thorium oxide.</u> http://tinyurl.com/kv74va8 - (Canada)
http://www.thehindubusinessline.com/economy/india-on-the-roadmap-of-tripling-nuclear-power-capacity/article9599683.ece

After our Department of Energy signed an agreement with China, we handed over all of our MSR data. To supply its needs while MSRs are being developed, China is relying on 27 conventional nuclear reactors plus 29 Generation III+ (solid fuel) nuclear plants that are under construction. China also intends to build an additional fifty-seven nuclear power plants, which is estimated to add at least 150 GigaWatts (GW) by 2030.

According to Reuters, Paris (June 28, 2016),

"Global increase in nuclear power capacity in 2015 hit 10.2 gigawatts, the highest growth in 25 years driven by construction of new nuclear plants mainly in **China**....

"We have never seen such an increase in nuclear capacity addition, mainly driven by China, South Korea and **Russia**,.. It shows that with the right policies, nuclear capacity can increase, said F. Birol, International Energy Agency's Executive Director, at a conference in Paris.

Dr. Alex Cannara - "When China National Nuclear Power Manufacturing Corporation. sought investors in 2015, they expected to raise a modest number of millions. They raised more than $280 billion."

http://dailycaller.com/2016/08/02/mit-china-is-beating-america-in-nuclear-energy/

In 2016, the Chinese Academy of Sciences allocated $1 billion to begin building **LFTRs** by 2020. As for Japan, which wisely began to restart its reactors in 2015, a **FUJI** design for a 100-200 MW LFTR is being developed by a consortium from Japan, the U. S. and Russia at an estimated energy cost of just three cents/kWh. Furthermore, it appears that five years for construction and about $3 billion per reactor will be routine in China.

https://www.technologyreview.com/s/602051/fail-safe-nuclear-power/

How a LFTR works

In one type of LFTR (there are several designs) a liquid uranium/salt mix circulates through the reactor core, releasing neutrons that convert Th-232 in an outer, shell-like "jacket" to Th-233. Thorium 232 cannot sustain a chain reaction, but it is *fertile*, meaning that it can be converted to *fissile* U-233 through neutron capture, also known as "breeding."

When the U-233 absorbs another neutron, it fissions (splits), releasing huge amounts of energy and more neutrons that activate more Th-232. In summary, a LFTR turns thorium into U-233, which thoroughly fissions, producing just a small amount of short-term waste in the process - about 10% as much as our conventional reactors currently produce.

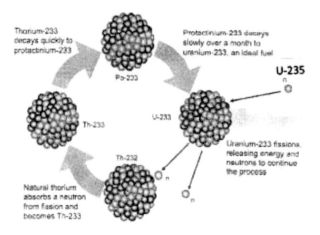

Neutron-induced breeding of thorium to fissile uranium

from *THORIUM: Energy Cheaper Than Coal* - Robert Hargraves

The half-life of Th-232, which constitutes most of the earth's thorium, is 14 billion years, so it is not hazardous due to its extremely slow decay.

Proliferation issues

It would be very difficult to make a weapon from LFTR fuels because the gamma rays emitted by the U-232 in the fuel would harm technicians and damage the bomb's electronics.

Uranium is most easily stolen during enriching, production of pellets, delivery to the reactor, and for long-term storage, but LFTRs only use external uranium to *start* the reaction, after which time uranium is produced *within* the reactor from thorium.

A conventional, 1 GW reactor requires about 1.2 tons of uranium oxide per year, but a 1 Gigawatt LFTR only requires a *one-time* "kickstart" of about 500 lbs. of U-235 plus about 1 ton of thorium per year during the reactor's 50 to 60-year lifespan.

Summary: Advantages of LFTRs

(Many of these also apply to MSRs that use Uranium.)

No CO2 emissions.

Not practical for making bombs. Proliferation *resistant.*

Produce only a small amount of low radioactivity waste that is benign in 350 years.

The liquid fuel, besides being at 700-1000 degrees C, contains isotopes fatal to saboteurs.

Do not require water cooling, so hydrogen and steam explosions are eliminated.

Don't require periodic refueling shutdowns because the fuel is supplied as needed and the by-products are constantly removed. (Conventional reactors are shut down every 18 – 24 months to replace about ¼ of the fuel rods, but, they can run much longer.)

Th-232 is far more abundant than U-235.

Well suited to areas where water is scarce.

Do not need huge containment domes because they operate at atmospheric pressure. Breed their own fuel.

Can't "melt down" because the fuel/coolant is already liquid, and the reactor can handle high temperatures.

Fluoride salts are less dangerous than the super-heated water used by conventional reactors.

Could replace the world's coal-powered generators by 2050. Are suitable for modular factory production, truck transport and on-site assembly.

Create Plutonium-238 that powers NASA's deep space exploration vehicles.

Are intrinsically safe: Overheating expands the fuel/salt, decreasing its density, which lowers the fission rate.

If there is a loss of electric power, the molten salt fuel quickly melts a freeze plug, automatically draining the fuel into a tank, where it cools and solidifies.

Are extremely efficient. At least 99% of a LFTR's thorium is consumed, compared to about 3% of the uranium in conventional reactors.

Are highly scalable - from 10 MW to 2,000 MW plants. A LFTR capable of generating 200 MW of electricity could be transported on a few semi-trailer trucks.

Cost much less than conventional reactors.

Robert Hargraves - *American Scientist* Vol. 98, July 2010. "Given the diminished scale of LFTRs, it seems reasonable to project that reactors of 100 megawatts can be factory produced for a cost of around $200 million."

Can't afford it?

While we temporize, Russia is building modular reactors (conventional and MSRs), for sale abroad, some of which will be mounted on barges that can be towed to coastal cities, thus making long transmission lines, with their 10% power loss, unnecessary.

In 2016, Russia inaugurated a commercial Fast Breeder Reactor (FBR) that extracts **nearly 100%** of the energy value of uranium. (Conventional reactors utilize less than 5%.) The FBR creates close to zero waste and guarantees that we will never run out of thorium, uranium and plutonium, which yield 1.7 million times more energy per kilogram than crude oil.

However, instead of pursuing these profitable programs, we have spent $400 billion on worthless F-37 jet fighters plus $2 billion PER WEEK in Afghanistan – AND there's that missing $8.5 *TRILLION* that the Pentagon can't find or explain.

Meanwhile, according to the GUARDIAN, "in 2013, coal, oil and gas companies spent $670 billion searching for more fossil fuels, investments that could be worthless if action on global warming slashes allowed emissions."

California plans a $50-100 billion high speed train to serve impatient commuters between San Francisco and Los Angeles, and in 2014, Wall Street shelled out over $28 billion in bonuses to needy executives. If you include greedy sports team owners and players who, between 2000 and 2010, received 12 billon tax dollars to help pay for their sports arenas, the total could exceed *$1 trillion*.

With that money, we could easily build enough MSRs to end the burning of fossil fuels for generating power while drastically cutting carbon dioxide production.

According to WORLD NUCLEAR NEWS, Russia's Rosatom Overseas intends to sell desalination facilities powered by nuclear power plants to its export markets:

"Dzhomart Aliyev, the head of Rosatom Overseas, says that the company sees 'a significant potential in foreign markets,' and is offering two AES-2006 LWRs producing 1200 MW each to Egypt's Ministry of Electricity and Renewable Energy as part of a combined power and desalination plant.

"Desalination units can produce 170,000 cubic meters of potable water per day with 850 MWh per day of electricity. This would use only about 3%

of the output of a 1200 MWe nuclear power plant. In addition, two desalination units are also being considered for inclusion in Iran's plan to expand the Bushehr power plant with Russian technology, and another agreement between Argentina and Russia also includes desalination with nuclear power."

In 2016, the Vice President of Rosatom reported that the company plans to build more than 90 plants in the pipeline worth some $110 Billion, with the aim of delivering 1000 GW by 2050.

Vladimir Putin – "by 2030 we must build 28 nuclear power units. This is nearly the same as the number of units made, or commissioned, over the entire Soviet period... ROSATOM, the Russian nuclear power corporation and builders of the Kundamkulam nuclear power plant in India, has orders for building many nuclear power units abroad."

Stratfor Global Intelligence reported in an October, 2015 article titled *Russia: Exporting Influence, One Nuclear Reactor at a Time* that "Rosatom estimated that the value of orders has reached $300 billion, with 30 plants in 12 countries. From South Africa to Argentina to Vietnam to… Saudi Arabia, there appears to be no region where Russia does not seek to send its nuclear exports."

However, in the U. S., our nuclear industry, opposed by Climate deniers like Donald J Trump, fervent "greens" and powerful companies that put profit before planet struggles to stay alive.

In *Why Not Nuclear?* **Brian King** described our failure to build Generation IV nuclear plants that, unlike LWRs take advantage of high-temperature coolants such as liquid metals or liquid salts that improve efficiency.

"Argonne National Laboratory held the major responsibility for developing nuclear power in the U.S. By 1980, there were two main goals: "Develop a nuclear plant that can't melt down, then build a reactor that can run on waste from nuclear power plants...

"In the early 80's Argonne opened a site for a Generation IV project in Idaho, and about five years later, they were ready for a demonstration. Scientists from around the globe were invited to watch what would happen if there was a loss of coolant to the reactor, a condition similar to the event at Fukushima where the cores of three reactors overheated and melted. Dr. C. Till, the director of the Generation IV project, calmly watched the gauges on the panel as core temperature briefly increased, then rapidly dropped as the reactor shut down without any intervention!

"The Argonne Generation IV project was a success, but it couldn't get past the anti-nuke politics of the 90s, so it was shut down by the Clinton administration because they said *we didn't need it.*

"One can only imagine what the world would look like today, with a fleet of Generation IV nuclear plants that would run safely for centuries on all the waste at storage sites around the globe. No CO_2 would have been created – only ever increasing amounts of clean, reliable power. So, why not nuclear power?

"Unfortunately, most environmentalists oppose nuclear power, as do many liberals. The Democratic Party is afraid of anti-nuclear sentiment... like the Nation Magazine, the Sierra Club and others. Why are all these people against such a safe and promising source of energy?

"... nuclear power has been tarred with the same brush as nuclear weapons. Nuclear power plants can't explode like bombs, but people still think that way....

"There is also a matter of group prejudice, not unlike a fervently religious group or an audience at a sports event of great importance to local fans. People are afraid to go against the beliefs of their peers, no matter how unsubstantiated those beliefs may be."

Now, some good news: In September, 2016, New York state adopted a clean energy policy that includes a guaranteed income clause that will benefit nuclear power.

PRO-NUCLEAR LINKS

http://thoriumforum.com/nobel-physicist-carlo-rubbia-thorium-trumps-all-fuels-energy-source

http://www.cnn.com/2013/11/03/world/nuclear-energy-climate-change-scientists-letter/

http://www.bbc.co.uk/news/science-environment-24638816

www.wano.info//article.cfm?id=the-human-cost-of-energy

Chapter 9

Blowin' Wind

When the City Council of Virginia, Minnesota - my hometown - lamented the pathetic return from the city's solar panels, they were right on target, but solar panels shine when compared to windmills.

I was thrilled when the first windmills appeared on a ridge north of town, but a few years later, having noticed a significant amount of "down time," I checked on wind power's record with the help of associates at the Thorium Energy Alliance and discovered that the windmill industry had been selling a lot more sizzle than steak.

During the "green" search for energy alternatives, which was guided by an "anything but nuclear" bias, the Sierra Club and several others *to which I once belonged* took pains to define what was "renewable" and what was not. In so doing, they deliberately excluded nuclear power, even though we have enough uranium and thorium to last 100,000 years. And because those who would profit from wind and solar said nothing about carbon footprints, environmental damage, resource use, inefficiency, bird, bat and human deaths (death prints) and the need for huge subsidies, we drank their Kool-Aid, and now wonder why it's making us sick. Well, here's why, from many points of view.

 # 1. Safety - <u>In the US alone</u>, windmills kill 1 million birds and 1 million bats every year, even as insect borne diseases like Zika, dengue fever and malaria are increasing. How green is that? Shouldn't environmentalists that, Save the Eagles International, "windmills kill 30 million birds and 50 million bats per year." Shouldn't they be concerned that Pacific Corp., which operates 13 wind farms, has sued the U S Interior Department to keep it from revealing how many birds and bats their windmills have killed?

Don't these "environmentalists" care that, according to the May, 2011 issue of Science magazine, a "single colony of 150 brown bats has been estimated to eat nearly 1.3 million disease-carrying insects each year"? Shouldn't they know that, according to the US Geological Survey, bats consume harmful pests that feed on crops, providing about **$23 billion** in benefits to America's agricultural industry every year?

And it's not just birds and bats. According to the *Caithness Windfarm Information Forum,* "Just in England, there were 163 wind turbine accidents that killed 14 people in 2011, which translates to about 1000 deaths per billion kilowatt-hours.

"In contrast, during 2011 nuclear energy produced 90 billion kWhrs in England with <u>NO deaths</u>. In that same year, America produced 800 billion kWhrs via nuclear with <u>NO deaths</u>."

Why is it almost sacrilegious for the Sierra Club and its clones to rethink windmills, and why do they refuse to watch presentations that compare the records of their supposedly green alternative energy sources to the record of CO_2-free nuclear power? Could $$$ be involved?

Researchers at the Imperial College London and the University of Edinburgh report that 117 of world's 200,000 windmills burn every year - far more than the 12 reported by wind farm companies – and even more throw their blades.

Why hasn't our media featured this image of two Dutch engineers waiting to die. One jumped to his death. The other burned to death. (It's been available for years.)

Source – Imgur

http://www.caithnesswindfarms.co.uk/AccidentStatistics.htm

Why doesn't the media publish any of these <u>easily available</u> images of burning windmills, windmills that have simply toppled over and windmills that have thrown their blades more than a third of a mile?
http://www.windaction.org/posts/38949-dual-deaths-in-wind-turbine-fire-highlight-hazards#.WD2uLWfrt9C

Source - Imgur

Insurance claims in the U.S. for 2012 show that blade damage and gearbox failures cost the industry $240,000 and $380,000 respectively. Claims associated with windmill foundations have averaged $1,300,000 per year, and they reached $2,500,000 in 2102 due to extreme circumstances.

For some example of the opposition we face from many environmentalists, please read Paul Lorenzini's excellent article titled "Saving the Environment from Environmentalism" at http://atomicinsights.com/saving-the-environment-from-environmentalism-2/ and https://www.youtube.com/watch?v=uqZTsy3Dav8

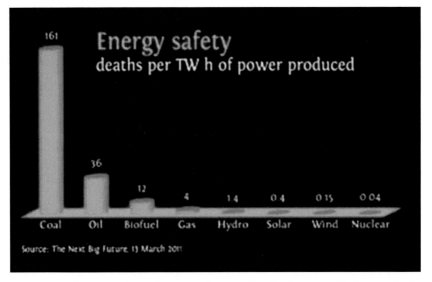

As mentioned near the end of chapter 7 (repeated here for emphasis) this chart reveals that if we give nuclear power a score of 1 on electricity generated per fatality – including Chernobyl - wind is 4 times worse, solar gets an 11, oil a 900, and coal comes in at 4,000.

Nuclear power is even safer than benign hydropower, which has a huge carbon footprint because of the energy needed to manufacture the cement in its concrete, and because reservoirs create large amounts of methane. (See *Hydro's Dirty Secret Revealed* by Duncan Graham-Rowe.)

Furthermore, people who are forced to live close to windmills have complained of severe sleep deprivation, chronic stress, dizziness and vertigo caused by low frequency noise and inaudible noise below 20 Hz, known as infrasound.

However, despite these problems, wind promoters have no interest in replacing coal-burners with highly efficient, ultra-safe, Generation III+ reactors or Molten Salt Reactors that *cannot melt down* and don't generate the hydrogen that exploded at Chernobyl and Fukushima.

With these facts in mind, how can "environmentalists" support the creation of intermittent wind farms that require carbon-burning backup generators, have large carbon footprints, only a 20-year lifespan, are difficult to recycle and have large death prints. Nuclear power, by contrast, operates 24/7, has a much smaller carbon footprint, a 60-year lifespan, is 90% efficient, requires very little land, and kills no birds or bats.

#2. Tilted Economics - I understand why power companies cooperated with the rush to wind power. For one thing, renewables were demanded by a misinformed public led by many of the "green" organizations whose *goals* I support, but not their methods.

For another, those 30% efficient windmills have received subsidies of $56.00 per megawatt hour. In comparison, 90% efficient nuclear power, which critics say is "too expensive," receives just $3.00 per megawatt hour.

Even the wind companies and **Warren Buffett** admit that without the subsidies, they'd be losers: "...on wind energy, we get a tax credit if we build a lot of wind farms. That's the only reason to build them. They don't make sense without the tax credit." (2014)

http://tinyurl.com/meule2r

True Cost of Wind Power – **Newsweek** – 4/11/15

"As consumers, we pay for electricity twice: once through our monthly electricity bill and a second time through taxes that finance **massive subsidies** for inefficient wind and other energy producers.

"Most cost estimates for wind power disregard the heavy burden of these subsidies on US taxpayers. But if Americans realized the full cost of generating energy from wind power, they would be less willing to foot the bill – because it's more than most people think.

"Over the past 35 years, wind energy – which supplied just 4.4% of US electricity in 2014 – has received U S $30 billion in federal subsidies and various grants. These subsidies shield people from the truth of just how much wind power actually costs and transfer money from average taxpayers to wealthy wind farm owners, many of which are units of foreign companies...."

https://blogs.spectator.co.uk/2017/04/flawed-thinking-heart-lethal-renewable-energy-swindle/#

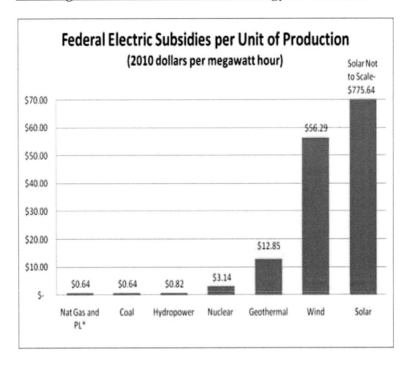

Testimony of **Dr. James Hansen**, formerly of NASA, to the Senate Foreign Relations committee March, 2014.

"Nuclear's production tax credit (PTC) of 1.8 cents/kWhr is *not indexed* for inflation. PTCs for other low carbon energies <u>are</u> *indexed*. The PTC for wind is 2.3 cents/kwhr.

"Plants must be placed in service before January 1, 2021. Thanks to Nuclear Regulatory Comm. slowness, that practically eliminates any PTC for new nuclear power.

"Do you know about "renewable portfolio standards"? If government cares about young people and nature, why are these not "**carbon-free** portfolio standards"?

"This is a huge hidden subsidy, reaped by only renewables. There is a complex array of financial incentives for renewables. Incentives include the possibility of a 30% investment tax credit in lieu of the PTC, which provides a large "time-value-of-money" advantage over a PTC spread over 8-10 years, accelerated 5-year depreciation, state and local tax incentives, loan guarantees with federal appropriation for the "credit subsidy cost.

"Nuclear power, in contrast, must pay the full cost of a Nuclear Regulatory Commission license review, at a current rate of $272 per professional staff hour, with no limit on the number of review hours. The cost is at least $100-200 million. The NRC takes a minimum of 42 months for its review, and the uncertainty in the length of that review period is a major disincentive."

From *Clean Technica* – October, 2015

"When supply is high and demand is low, spot prices generally fall — this is especially true in markets with high shares of renewable energy. What precipitates *negative* pricing are conditions which encourage energy producers to sell at an apparent loss, knowing that in the *longer* term [thanks largely to huge taxpayer subsidies] they will still profit.

"The Texas grid is called ERCOT, and it's managed by the energy agency of the same name... The market functions through auctions, where energy producers place a competitively priced bid to supply some amount of energy at a particular time and particular price...

"Various subsidies. including U. S. federal production tax credits and state renewable energy certificates, compensate wind power producers... to such an extent that it allows wind farms to continue to make money even when selling at negative prices."

We are all paying hidden costs to prop up these inefficient, deadly "alternatives" that depend on methane to produce 70% of their rated power, even though the methane leakage from fracking and the distribution system are erasing any benefits we have hoped to get by avoiding coal. Furthermore, the price quoted for a nuclear plant includes the cost of decommissioning, but it isn't for the thousands of windmills or solar farms that only last 20 years.

In fact, the deck has been stacked against nuclear power by "green" profiteers and carbon lobbyists who know they cannot compete with 90% efficient, CO2-free nuclear power. Still, despite the bureaucratic handicaps on nuclear power and the advantages given to renewables, nuclear power is financially competitive, as the following chart clearly reveals.

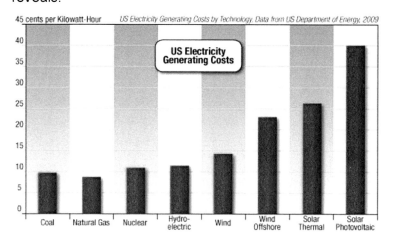

#3. Misrepresentation and inefficiency - When wind advocates promote the glories of wind power, they use numbers based on the windmill's nameplate rating, its *maximum* capacity – as in a February 20, 2015 Earth Watch article, which said, "...the total amount of wind power *available...* has grown to 318,137 megawatts in 2013."

But because wind power is highly intermittent, windfarms usually generate an average output of about 30% of their capacity, which is why 318,137 megawatts is very misleading, and 90,000 would be more accurate, probably even generous. When they say that windmills can supply xxxxxxx homes, they are usually talking about the plate rating on the generator – the output under ideal conditions, not the average amount of power they really produce.

https://www.eia.gov/electricity/monthly/epm_table_grapher.cfm?t=e pmt_6_07_b

Neither solar nor wind can deliver 24/7 "baseload" power – the 90% of our power that is provided by nuclear plants plus hydropower, natural gas, oil and coal. Of those five, only nuclear power plants (despite Chernobyl, a plant deemed to be "illegal" everywhere else in the world), has been safely delivering carbon dioxide-free power for more than fifty years.

Britain, faced with the choice of building 12 nuclear plants or the 30,000 1-MW windmills needed to provide an equal amount of power, chose nuclear. And Japan, which closed its nuclear plants in a burst of post-Fukushima panic, has begun to reactivate them, which will reduce the thousands of tons of CO2 they've been dumping into our already polluted atmosphere by burning methane.

Germany, which over-reacted by closing its nuclear plants in favor of wind and solar, is paying almost four times more for power than nuclear France. And with its industries hurting, the Merkel government has begun to rethink nuclear power. While they debate, they are creating more CO2 by burning lignite, the dirtiest member of the coal family.

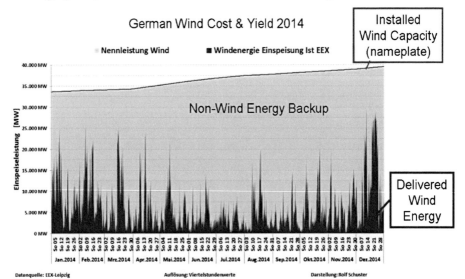

Germany paid for the top line of this graph, but only got the dark blue spikes. The light blue area above the spikes is mostly supplied by burning some form of carbon, which worsens Climate Change. (Every megawatt of wind generation capacity requires at least another MW of natural gas or coal generation for backup.)

http://www.thegwpf.com/germany-faced-huge-cost-of-wind-farm-decommissioning/

https://parkergallantenergyperspectivesblog.wordpress.com/2016/12/06/how-much-is-wind-power-really-costing-ontario/ (31 cents / kwh)

http://issues.org/33-2/inside-the-energiewende-policy-and-complexity-in-the-german-utility-industry/

#4. Methane - Windmills generate no more than 30% of their rated capacity, so a windmill with a name plate rating of 1 megawatt will actually average only about 0.3 mW, which means that the missing 70 % has to be supplied by power plants that, for the most part, burn carbon - usually coal or natural gas - which is largely methane, which produces more CO_2. I repeat, for emphasis, that methane, over its lifetime, is twenty times worse than carbon dioxide as a greenhouse gas, but during its youth, it is at least 75 times worse than CO_2 - and the next ten to twenty years are years of deep concern. Unfortunately, there's more!

Recent ground and aerial surveys reveal that huge amounts of methane are leaking from fracking wells, storage facilities and our distribution systems, and this fugitive methane is *more than* offsetting any gains we might have made by burning natural gas instead of coal. In other words, our reliance on bird, bat and human-killing windmills has been a step backward for wildlife *and* the environment.

In Boston, ground-based measurements suggest that methane leaks are everywhere.

While we pollute our aquifers by fracking for methane to assist inefficient windmills, we are simultaneously flaring (burning) huge volumes of natural gas across much of North Dakota and elsewhere because it's "too costly" to pipe it to market. In short, we are wasting it "here" while we frack for more "there." As usual, it's all about $$$. How green is that?

Windmills, in effect, are glorified, heavily subsidized carbon burners that needlessly create more of the carbon dioxide that we seek to avoid. Were it not for our misguided passion for inefficient renewables, we'd have less need for fracking and less of the environmental damage it produces.

In 2014, satellite images of oil and gas basins in Texas and North Dakota revealed staggering 9 -10% leakage rates of heat-trapping methane. Because of these leaks, fracking accelerates climate change even *before* the methane it extracts is burned.

A year later, thanks to a "discovered" email message from Lenny Bernstein, a thirty-year oil industry veteran and ExxonMobil's former in-house climate expert, we learned that

Exxon accepted the reality of climate change in 1981, long before it became a public issue - but then, Exxon spent at least $30 million on decades of climate change denial.

In addition, despite studies from Johns Hopkins that reveal an association between fracking and premature births, high-risk pregnancies and asthma, Pennsylvania health workers were told by their Department of Health to ignore inquiries that used fracking "buzzwords."

http://www.truth-out.org/news/item/38022-where-has-the-waste-gone-fracking-results-in-illegal-dumping-of-radioactive-toxins

According to a 2014 U N report, atmospheric methane levels have never exceeded 700 parts per billion in the last 400,000 years, but they reached 1850 ppb by 2013.

http://energyrealityproject.com/lets-run-the-numbers-nuclear-energy-vs-wind-and-solar/

In 2015, a Duke University study reported: "Thousands of oil and gas industry wastewater spills in North Dakota have caused "widespread" contamination by radioactive materials, heavy metals and corrosive salts, putting the health of people and wildlife at risk."

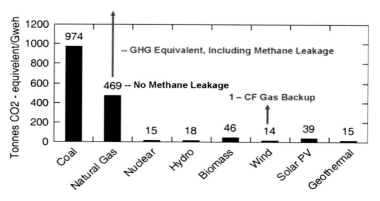

Courtesy Burton Richter — *Comparison of Life Cycle Emissions in Metric Tonnes of CO_2e per GW-hour for various modes of Electricity Production; P.J. Meier, Life-Cycle Assessment of electricity Generation Systems with Applications for Climate Change Policy Analysis,*

In their excellent *Wind and Solar's Achilles Heel*: The Methane Meltdown at Porter Ranch, **Mike Conley** and **Tim Maloney** reported:

"Even a tiny methane leak can make a gas-backed wind or solar farm just as bad – or worse – than a coal plant when it comes to global warming. And the leaks don't just come from operating wells. They can happen anywhere in the infrastructure... In the U.S., these fugitive methane leaks can range up to 9%.

"If the fugitive methane rate of the infrastructure... exceeds 3.8%, then you might as well burn coal for all the "good" it'll do you. All in all, the numbers are pathetic - some of the most recent measurements of fugitive methane in the U.S. are up to 9%. But the gas industry predictably reports a low 1.6%."

The sediments in many of the world's shallow oceans and lakes also release vast amounts of methane from frozen organic matter as it thaws and decomposes. When a Russian scientist searched the Arctic shores for methane, he found hundreds of yard-wide craters, but when he returned a few years later, they were 100 yards in diameter.

http://www.unep.org/newscentre/default.aspx?DocumentID=2698&ArticleID=9338#sthash.MNhnIIkM.dpuf> report,

In 2014, **N. Nadir,** of the Energy Collective wrote,

"The most serious *environmental* problem that renewable energy has is that even if it reached 50% capacity *somewhere,* this extraordinary waste of money and resources would still be dependent on natural gas, which any serious environmentalist with a long-term view sees as disastrous.

"Natural gas is not safe - even if we ignore the frequent news when a gas line blows up, killing people. It is not clean, since there is no place to dump its CO2; it is not sustainable; and the practice of mining it - fracking - is a crime against all future generations who will need to live with shattered, metal-leaching rock beneath their feet, and huge amounts of CO2 in the atmosphere."
http://theenergycollective.com/jemillerep/450556/what-are-capacity-factor-impacts-new-installed-renewable-power-generation-capaciti

Tim Maloney of the Thorium Energy Alliance argues that we should be *conserving* natural gas because methane is the primary feed stock for ammonia, and ammonia is required to produce nitrogen-based fertilizers, a shortage of which could cause starvation. In addition, closing nuclear plants and expanding "renewables" that require natural gas will substantially increase CO2 and methane emissions.

#5. Longevity and Reliability - Because 30% efficient windmills only have 20-year lifespans (at best), they must be rebuilt two times after initial construction to match the 60-year lifespan of 90% efficient nuclear power plants.

Here's what an anonymous wind technician from Bismarck, North Dakota, had to say about the usefulness of windmills:

"Yeah, we all want to think we're making a difference, but we know it's bullshit. If it's too windy, they run like sh__, if it's too hot, they run like sh__, too cold, they run like sh__. I just checked the forecast, and it's supposed to be calm this weekend so hopefully not very many will break down, but hell man, they break even when they aren't running. I've given up on the idea that what I'm doing makes a difference in the big picture. Wind just isn't good enough."

http://nordic.businessinsider.com/after-all-the-money-poured-into-wind-energy-denmark-admits-its-too-expensive-2016-5/

#6. Resources and materials - Organizations like the Sierra Club wear blinders that exclude their defects, and when I or my associates offer presentations on the safety records and costs of the various forms of power generation, including nuclear, we rarely get a reply, and my Minnesota chapter is a case in point.

Because of those blinders, they apparently don't know or care that It will take 9,500 1-MW windmills running their entire life spans to equal the life-cycle output of just one nuclear plant. Perhaps they don't realize that those windmills, which last just 18-20 years, require far more steel and concrete than just one nuclear plant that is *good for 60 years.* As a consequence, the "carbon footprint" of those 30% efficient windmills is much larger than that of a 90% efficient nuclear power plant.

Dr. Alex Cannara:

"The material in five, 2 MW windmills (10 MW total) could build a complete 1 GW nuclear power plant that will generate *~100x the power*, on 1/1000 the acreage, with no threat to species or climate."

http://www.aei.org/publication/wind-an-even-bigger-boondoggle-than-ethanol/

https://www.masterresource.org/windpower-problems/wind-power-least-sustainable-resource/

http://www.wbur.org/hereandnow/2016/12/23/aging-wind-

Furthermore, the wind industry doesn't know what to do with these 170- foot, 22,000 pound, fiberglass blades that last less than 20 years and are so difficult to recycle that many landfills won't take them.

A 1-GW wind farm needs 1300 tons of new blades per year, and because they cost $100k each, that's $200 million every 18 years, or $33.6 million per year per gigawatt created just for the blades. All this for a rig that requires carbon-burning generators to supply the 70% of their rated power that they don't supply.

Those who guide the Sierra Club or Greenpeace, etc., should know that most windmills require magnets made from rare earth minerals like neodymium that are primarily mined in China where mining and refining the ore has created immense toxic dumps and lakes that are causing high rates of skin and respiratory diseases plus cancer and osteoporosis. If they know this, why are they silent? If they don't, they should.

Please do a computer search for "Baotou, China".

http://www.bbc.com/future/story/20150402-the-worst-place-on-earth

According to the Bulletin of Atomic Sciences, "a two-Megawatt windmill contains about 800 pounds of neodymium and 130 pounds of dysprosium."

Unlike windmill generators, ground-based generators use electromagnets, which are much heavier than permanent magnets, and do not contain any rare-earth elements.

That might seem dull, but here's the problem: Accessing just those two elements produces tons of waste that includes arsenic and other dangerous chemicals. And because the U.S. added about 13,000 MW of wind generating capacity in 2012, that means that some 5.5 million pounds of rare earths were refined just for windmills, which created 2,800 tons of toxic waste.

For perspective, our nuclear industry, while creating 20% of our electricity, produces only about 2.35 tons of spent nuclear fuel (commonly called "waste"), per year, which they strictly contain, but the wind industry, while creating just 3.5% of our electricity, is making much more radioactive waste where rare-earths are being mined and processed – and its disposal is virtually unrestricted.

We know that it takes several thousand windmills to equal the output of one run-of-the-mill nuclear reactor, but to be more specific, let's tally up all of the materials that will be needed to replace the closed Vermont Yankee nuclear plant with renewables.

Dr. Tim Maloney has done just that, writing,

"Here are numbers for wind and solar replacement of Vermont Yankee:

1) Amount of steel required to build wind and solar;
2) Concrete requirement;
3) CO_2 emitted in making that steel and concrete;

4) Money spent;
5) Land taken out of crop production or habitat.

Let's assume a 50/50 split between wind and solar, and for the solar a 50/50 split of photovoltaic (PV) and CSP concentrated solar power. (mirrors.)

To replace Vermont Yankee's 620 MW, we'll need 310 MW (average) for wind, 155 MW (avg) for PV solar, and 155 MW (avg) for CSP... Using solar and wind would require:

Steel: 450,000 tons; that's 0.6% of our U.S. total annual production, just to replace one smallish plant.
Concrete: 1.4 million tons; 0.2% of our production/yr.
CO_2 emitted: 2.5 million tons
Cost: about 12 Billion dollars
Land: 73 square miles, which is larger than Washington DC, just to replace one smallish nuclear plant with solar/wind….

ALTERNATIVES

a. Replace Vermont Yankee with a Westinghouse /Toshiba model AP1000 that produces 1070 MW baseload, ~ <u>2 x the output of Yankee.</u>

Normalizing 1070 MW to Vermont Yankee's 620 MW, the AP1000 uses:

Steel: 5800 tons – 1 % as much as wind and solar.
Concrete: 93,000 tons – about 7% as much.
CO_2 emitted: 115,000 tons [from making the concrete and steel] - about 5% as much.
Cost: We won't know until the Chinese finish their units. But it should be less than our "levelized" cost. [Perhaps $4-5 billion]

Land: The AP1000 reactor needs **less than ¼ square mile** for the plant site. Smaller than CSP by a factor of 2000. Smaller than PV by a factor of 4,000. Smaller than wind by 13,000.

b) Better yet, we could get on the thorium energy bandwagon. Thorium units will beat even the AP1000 by wide margins in all 5 aspects – steel, concrete, CO2, dollar cost, and land." tinyurl.com/lsvq2yn

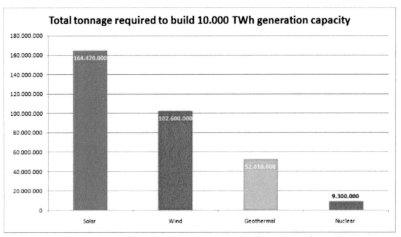

Source : Quadrennial technology review / Capter 10 : Concepts in integrated analysis | September 2015 | www.energy.gov

https://www.vestas.com/~/media/vestas/about/sustainability/pdfs/lca_v112_study_report_2011.pdf

In *How to decarbonize? Look at Sweden,* (Bulletin of the Atomic Scientists), we are told, "To light the way forward, we need to examine success stories where nations have greatly reduced their carbon dioxide emissions while maintaining vigorous growth in their standard of living: a prime example is Sweden. Through a combination of sensible government policies and free-market incentives,

Sweden has managed to cut its per capita emissions by a factor of 3 since the 1970s, while doubling its per capita income and providing a wide range of social benefits. They did this by building nine nuclear reactors."

Fortunately, countries not plagued by fear-mongers are expanding nuclear power countries like South Korea, Sweden, India and Russia, which is even *exporting* reactors. Now add China, which, in addition to its 27 current nuclear plants, has 29 ultra-modern plants under construction with plans to build 57 more.

These nations have let science guide their decisions – not the hoopla produced by windmill profiteers or the opposition of well-meaning greens who have closed their minds to science. That science clearly reveals that these pretty, white windmills should be painted 1/3 red for the birds, bats and humans they kill, 1/3 black for the carbon we must burn when they're (mostly) not working and 1/3 gold for the subsidies – the tax $$$$ - they consume.

https://blogs.spectator.co.uk/2017/04/flawed-thinking-heart-lethal-renewable-energy-swindle/#

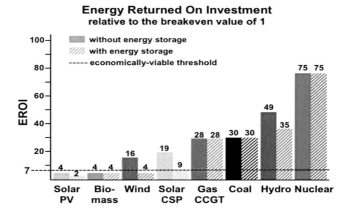

CHAPTER 10

Solar and Biomass

Concentrated Solar Power – CSP

"Man has lost the ability to foresee and forestall—he will end by destroying the earth." Albert Schweitzer

https://davidgattieblog.wordpress.com/2016/07/25/overselling-california-solar/

http://energyfairness.org/trouble-at-ivanpah-silence-from-sierra/

Built with a $1.6 billion federal loan guarantee and the support of the Sierra Club, California's bird-broiling Ivanpah facility uses 350,000 mirrors to focus sunlight onto towers in which fluids are heated to 1,000 degrees F. However, the facility only delivers 23% of its rated power, the rest of which is provided by burning carbon. (Birds that fly too close to the towers are "incinerated" in mid-flight.)

Like windmills, CSPs are carbon burners due to their 30% efficiency and their need to "heat things up" with natural gas every day before sunrise. And since 2013, Ivanpah's owners have twice sought permission to use even more gas than was allowed under the plant's certification. (1.4 Billion cubic feet in 2016)

Since 2000, Spain has paid renewable corporations $41 billion *more* for their power than it received from consumers, so, not surprisingly, in 2015, the government finally decided to slash subsidies to solar power, especially CSP. In 2013, solar investment in Spain dropped by 90 percent from its 2011 level, and worldwide interest in CSP is falling fast.

Photovoltaic solar – PVs

During 2014 - 2016, we produced about 3,500,000 PV panels per year. Copper, aluminum, high-quality quartz and rare earth materials are required to manufacture these panels – and if we were to try to get just half of our power from solar panels, we'd need <u>billions</u> of them.

Although PVs share most of wind's defects, PVs are less hostile to wildlife than windmills. However, because solar panels wear out in just two decades (like windmills), we will periodically need to mine more materials and recycle billions of them, which will require more energy. In the ensuing process thousands of tons of toxic by-products and additional CO_2 will *again* be created. PVs, like wind power, should be limited to bringing electricity to remote communities that are far from the grid.

Germany

Thanks to our biased, science-ignorant media, we've all read that "Germany gets half of its energy from solar panels." That might be true a long, sunny, mid-summer day but in reality, Germany's <u>official</u> statistics reveal that the correct figure for long-term production is ten times lower, only 4.5%.

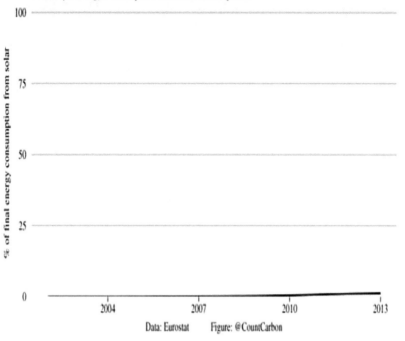

Because of Germany's knee-jerk response to Fukushima, 46 percent of their electricity now comes from biomass and coal, half of which is lignite, which, I repeat, is the dirtiest member of the coal family.

As a consequence, Germany's CO_2 levels are soaring, and many consumers are becoming energy poor due to rising electricity prices and taxes that subsidize their "green" energy.

Up to 800,000 Germans have had their power shut off because they couldn't pay their bills. In addition, building the 17,000 miles of power lines (which can lose 10% of the power), to serve Germany's renewables is expected to cost $27 billion. Some manufacturers, faced with rising power bills are heading to the US, where power prices are 1/3 of Germany's.

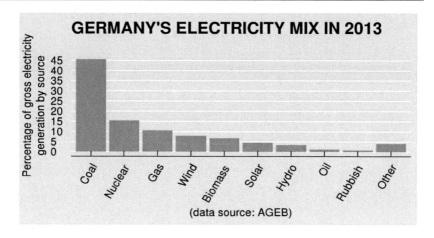

These statistics verify that Germany's "solar revolution" is mostly wishful thinking.

Climate news from Germany in English
by Pierre L. Gosselin   January 2015

"Former German Economics Minister Wolfgang Clement says that Germany's once highly ballyhooed transition to green energy *'has careened out of control'* and has hurt the country economically. He also says that the naivete' involved in implementing green energies has been 'breathtaking' and has turned into 'a disaster'.

"Germany's Energiewende has been criticized as the main driver behind the country's high electricity prices, unstable power grid conditions, growing energy poverty and for marring the landscape with inefficient and ugly industrial wind turbines.

"According to Minister Clement, Germany's electricity prices are among the highest in Europe and have led energy-intensive heavy industries to pack up and leave."

United Kingdom

As of 2015, British consumers pay more than $1.66 billion a year in subsidies to renewable energy producers. As in Germany, about 18% of the nation's population is in energy poverty due to high energy prices and subsidies for alternative sources like windmills, which must be expensively overhauled every **20 years.**

Denmark

Denmark has been heading the vanguard in the battle for windpower, but now admits it's too expensive - Reuters

Karl-Johan Byttner, May, 2016

"In 2015, Denmark set a new world record by generating the equivalent of 42.1 percent of the country's total energy consumption by wind. Denmark is also the world's largest exporter of wind power equipment, so it's probably fair to say that Denmark is perhaps the world's leading wind power nation....

"In 2016, the Danish government decided to abort the plans to build five offshore wind power farms, which were to stand ready by 2020. At the same time, Denmark is also scrapping its green energy tariffs and abandoning some of its climate goals.

"Since 2012 when we reached the political agreement, the cost of our renewable policy has increased dramatically," said Minister for Energy and Climate Lars Christian Lilleholt

"The cost of subsidizing wind power has become heavier as energy prices in the Nordic countries [hydroelectric Norway and nuclear Sweden] have fallen dramatically, making the renewable alternatives less attractive.

"The Danish consumers and companies pay the highest prices for electricity within the European Union, according to the European Electricity Association.

"The analysis showed that in 2014 a staggering 66% of the average Danish electricity bill went to taxes and fees, 18% to transportation and only 15% of the price for the electricity but Germans paid 52% in electricity taxes."

U A E

Although the United Arab Emirates has some of the best solar resources in the world, they have decided to spend $20 billion on nuclear reactors because they can operate for 75 - 80 years at 90 % efficiency instead of installing 30% efficient solar farms with 20-year lifespans.

U. S.

In 2015, our nuclear plants created 839 terawatt-hours of CO2-free electricity. That's four times as much as all CO2-creating wind projects, 21 times as much as all CO2-producing U.S. solar, and three times as much as all U.S. hydropower facilities. And in 2016, the National Academy of Sciences reported that the cost of subsidies for 30% "carbon-free" renewables like wind and solar is a stunning $250 for each ton of CO2 saved. Unfortunately, these "alternative" energy sources tend to displace highly reliable nuclear plants that, paradoxically, get no compensation for being CO2-free. http://www.nationalreview.com/article/438038/nuclear-power-necessary-green

Dr. Alex Cannara:

"Half a billion PV panels [as proposed by Sec. Hillary Clinton] will add about 800,000,000 kW of unnatural global warming when the sun is out [because those dark panels get hot]. This is equivalent to building about 5,000,000 new homes with black roofs in sunny climes, or adding about ten million gasoline/diesel vehicles to the road.

"The Topaz PV facility in California, which cost $2.5 billion, requires 9 square miles of panels to produce an average of 250MW. That's just 0.043 MW/acre.

"In contrast, Arizona's Palo Verde nuclear plant cost $5.9 billion, produces 3,900 MW 24/7 for just $0.03/kWHr. That's 1MW/acre, so the nuclear plant generates 25 times more power per acre."

http://wattsupwiththat.com/2015/07/29/the-green-mirage/

http://canadianenergyissues.com/2017/03/10/why-the-ontario-government-continues-to-endorse-and-make-ratepaers-cover-bad-cheques/

Rebuilding the Power Grid to Handle Solar and Wind is Absurdly Expensive

The Daily Caller News Foundation – Andrew Follett

"The three power grids that supply the United States with energy are massive and expensive pieces of infrastructure. The grids are valued at trillions of dollars and can't be replaced in a timely manner. It takes at least a year to make a new transformer, and they aren't interchangeable, because each unit must be built specifically for its location.

"At a time when the U.S. government is more than $18 trillion in debt, building power grids that can handle solar and wind may not be feasible.

"Merely building a 3,000-mile network of transmission lines capable of moving power from wind-rich West Texas to market in East Texas proved to be a $6.8 billion effort that began in 2008 and still isn't entirely finished.

"Building the infrastructure to move large amounts of solar or wind power from the best places to generate it to the places where power is needed could be incredibly expensive and could cost many times the price of generating the power."

http://oilprice.com/Energy/Energy-General/How-Intermittent-Renewables-Are-Harming-The-Electricity-Grid.html

https://carboncounter.wordpress.com/2015/06/04/why-wind-farms-can-be-relied-on-for-almost-zero-power

https://carboncounter.wordpress.com/2015/06/05/a-case-study-in-how-junk-science-is-used-by-anti-nuclear-environmentalists/

Acknowledged Subsidies

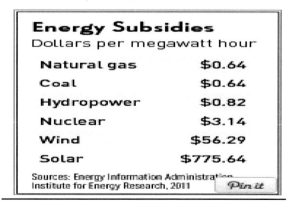

Hidden Subsidies

Besides selling solar energy for 4 to 5 cents/kWh, the operators of solar farms also sell solar renewable energy credits (SRECs), to companies like Apple that buy these credits for up to 40 cents/kWh to *greenwash* their images. SRECs are also auctioned to power companies that are required by state laws to buy enough to claim that x % of their power is from solar sources, which costs the utility and its customers another 30 cents/kWh.

http://www.srectrade.com/srec_markets/massachusetts
http://www.solarsystemsma.com/srecs.php.

These rules have created fertile ground for scams: One Vermont solar farm was able to sell electricity for thirty cents/kWh because the developer of the "farm" contributed to the campaigns of the politicians who passed the law that requires utilities to pay that price.

http://www.theenergycollective.com/sol-systems/2385104/the-moment-weve-been-waiting-for-massachusetts-srec-clearinghouse-auction-results

According to **Dr. Robert Hargraves**, "This is an enormous subsidy, paid for by the utility, which gets the money from its consumers. "Another example is home rooftop solar. The utility normally buys power at about 5 cents/kWh from generators, adds its costs, then sells it at about 15 cents/kWh to homeowners.

"With subsidies, when the owner's panels are generating power, the meter runs backwards, selling power to the utility for 15 cents/kWh, which they could have bought for 5 cents per kWh from their normal suppliers.

"Now this has transformed into a community solar scam, where multiple homeowners (presumably with north-facing roofs, or shade trees) can mount their solar panels in a community solar plan. Really, they've just become investors in a scam that benefits the solar industry."

And for those who proposed coating roads with PV cells, which many scientists opposed, The *DAILY CALLER NEWS FOUNDATION* (10-19-16) reported:

"Solar Road is 'Total and Epic' Failure"

"Despite massive internet hype, the prototype solar 'road' can't be driven on and hasn't generated any electricity. Roughly 25 out of 30 panels installed in a prototype solar road in Idaho broke within a week... Every single promise made about the prototype seems to have fallen flat and the project appears to be an epic failure, according to an electrical engineer.

"The U. S. Dept. of Transportation granted $750,000 to fund the research into the scheme, then invested another $850,000."

And remember, none of this would be needed if we hadn't been conned into avoiding nuclear power. (Crowd funding raised another $2.25 million bringing the total cost of this disaster to $3.9 million.)

The Guardian:

Dr. James Hansen is critical of Hillary Clinton's plan to put 500,000,000 solar panels on rooftops across the country: "You cannot solve the problem without a fundamental change, which means you have to make the price of fossil fuels **honest.**

"Subsidizing solar panels will not solve the problem... We have two political parties; neither wants to face reality. Conservatives pretend it's all a hoax, and <u>liberals propose solutions that are non-solutions."</u>

The American Humanist Assoc., <u>a liberal organization of which I am a former V P, unfortunately provides an example of liberal anti-nuclear bias and blind support of environmentally harmful "solutions".</u> The AHA should be a leader on combatting climate change and promoting nuclear power, but it has refused to print letters that dispute their support of inefficient, anti-environment, deadly renewables, and has also refused to update its biased, outdated position paper that opposes nuclear power. I am embarrassed to say that, on these issues, the AHA is well behind the Dali Lama and several religious organizations that include the Roman Catholic Church.

In 2015 and 2016, I tried to persuade the AHA board and the editor of the Humanist magazine to devote more time and space to climate issues and to update their position paper on nuclear power, but I was repeatedly ignored or rebuffed.

However, I was eventually asked to provide an updated position paper for the board to consider, which I and an associate subsequently submitted.

Unfortunately, their request was apparently just an empty gesture. Although I repeatedly asked if the board had discussed our revision, and what the result was (by email, so I would have a record), they never responded.

By down-rating climate change, refusing to rethink nuclear power and ignoring repeated requests for a progress report, the AHA has, in effect, joined the D J Trump and Caldicott crowd, and it has persuaded me to revise my will, which will cost them dearly.

The text that follows includes the essential portion of the lengthy AHA position paper on nuclear power (which is the result of LNT thinking unreasoned fear), and a proposed, July, 2015 suggested revision written by Chris Uhlik (Stanford PhD EE, Google Engineering Director and contributing designer / VP Engineer for Thorcon Power) and myself. (Our revision was vetted by nuclear physicists, engineers, experts in nuclear medicine and others with experience in nuclear power.)

The AHA Position Paper

WHEREAS cancer is on the increase throughout our population and has already become a major cause of disability and death; and...

WHEREAS radiation produced as byproducts of nuclear power plants, including plutonium, tend to produce widespread and long-lasting carcinogenic effects; and...

WHEREAS approximately one quarter ton of this plutonium is produced by such power plants, each with no proven way to eliminate its hazards permanently; and...

WHEREAS the particles of such plutonium and other nuclear waste byproducts tend to cause cancer when the human body inhales or otherwise comes in contact with it, even in very small quantities…

THEREFORE, be it resolved that no additional nuclear power plants be built anywhere in this country until safeguards have been provided to ensure against radiation hazards and for safe disposal of both low-level and high-level wastes, and that all present nuclear plants be phased out as soon as practical, until such time as adequate safeguards have been developed.

Suggested Revision
by Chris Uhlik and George Erickson

WHEREAS cancer is primarily an old age disease with many causes, including the thousands of untested chemicals in our environment, but is not caused by nuclear power plants, which are required by the Nuclear Regulatory Agency to emit less than 1/4 of the radiation that humans receive from their own internal radioactive emissions, as from food-borne Potassium-40, which is minimal in comparison to coal-burning power plant emissions that include radon, mercury, arsenic, uranium, cyanide and harmful particulates while exposing humans to 100 times more radiation than nuclear plants – which create no carbon dioxide, and   …

WHEREAS radiation and plutonium produced in nuclear power plants are either kept within the plant, where workers experience cancer rates no greater than that of the general population, or are stored at the plant or other licensed sites where it's impossible for the public to consume any of these materials, and …

WHEREAS the "waste" we store contains valuable isotopes that could be recycled – as the French do – and all of our waste created since 1960 would only cover one football field to a depth of 9 feet – of which only 10 % is hazardous if we recycle, and can be safely stored via deep borehole disposal, and...

WHEREAS there is no way that the public can inhale or otherwise come in contact with the elements used in nuclear reactors, including plutonium, and we suffer much greater dangers from fossil fuel emissions due to the avoidance of clean fission energy that cause occupational hazards...
http://www.msha.gov/stats/centurystats/coalstats.asp and

air pollution - see
https://www.google.com/search?q=beijing+air+pollution&safe=off&espv=2&biw=1663&bih=993&tbm=isch and

water pollution - see
https://www.google.com/search?q=beijing+air+pollution&safe=off&espv=2&biw=1663&bih=993&tbm=isch#safe=off&tbm=isch&q=coal+ash+spill and

land destruction – see
https://www.google.com/search?q=beijing+air+pollution&safe=off&espv=2&biw=1663&bih=993&tbm=isch#safe=off&tbm=isch&q=mountain+top+removal>

THEREFORE, be it resolved that CO2-free nuclear power, be judged by comparing its safety record, efficiency, ability to provide baseload power 24/7, facility lifespan, resource requirements, environmental consequences (including the destruction of natural habitat), to the records of the other means of power production, the largest of which create CO2 and, in the case of oil and natural gas wells, liberate vast amounts of methane, which is initially 75 times more damaging to the environment as CO2. We never received a response!

What About Biomass?

Biomass advocates claim that the carbon dioxide produced by burning biomass will be absorbed by forests, which supposedly makes it renewable – but that's ludicrous. When we burn fuel to level our forests, we create CO2 while leaving fewer trees to absorb the added CO2. Furthermore, wood-burning power plants, because of their low efficiency, emit about 50% more CO2 than coal per unit of energy produced.

Biomass smoke contains carcinogens like chromium, lead, nickel, benzene, toluene and formaldehyde, which explains why forest fire smoke proved fatal to 3 million people between 1996 and 2006. In addition, due to increasing European demand, wood pellet production is predicted to consume an additional 15 million acres of our forests within just a few years.

According to EuroStat, in 2013, biomass provided 64.2% of Europe's "renewable" power, which is absolutely outrageous.

A year later, the *DETROIT FREE PRESS* quoted a University of Michigan study by professor John DeCicco, who concluded that "Despite their purported advantages, biofuels from crops like corn or soybeans cause more CO2 emissions than gasoline. <u>We are "harvesting" trees that can individually absorb 10 pounds of air pollutants per year and create about 250 pounds of oxygen.</u>

This biomass must then be shipped to Europe, which creates even more CO2. Why, I ask, should we denude our forests to prop up Germany's "green" rush to inefficient alternatives – a rush powered by their foolish LNT-prompted mantra – *anything but nuclear power?*

As **Mathijs Beckers** wrote in SCIENCE A LA CARTE, With biomass, "... we've made our coal-fired power plants slightly less polluting, but even more destructive. Think about the sheer volume of coal-fired power plants all over the world that might be converted into these wood-eaters. In 2013, 40 billion pounds of wood pellets (shredded and pelletized trees) were burned for bioenergy.... This is the *green paradox*, to accept the destruction of natural cover under the guise of producing renewable energy

Trees awaiting transport to Germany, there to worsen Climate Change due to Angela Merkel's fear of nuclear power.
Image by Robert Surdey

However, there is a glimmer of light: Bloomberg News, in 2016, reported that some environmentalists have turned their backs on making ethanol from corn because of the program's many defects.

"'The big green groups that got invested in biofuels are tacitly realizing their blunder... It's really hard for people who hate oil, to think that this alternative that they have been promoting is even worse than oil,' said John DeCicco, a research professor at the University of Michigan Energy Institute who previously focused on automotive strategies at the Environmental Defense Fund."

Perhaps this glimmer of light will lead to the rejection of carbon-intensive wind and solar schemes, to the expansion of safe, efficient, CO_2-free nuclear energy, and to a huge increase in plug-in electric vehicles with regenerative braking, which reduces recharge needs by 10-15%.

Finally, I urge everyone to read Tim Maloney's excellent rebuttal of anti-nuclear "greens" who think we can satisfy our rising increasing energy needs with "alternative" sources like wind, water and solar (wws) alone.

For the facts on the EU's failed 1 Trillion Euro bet on renewables see http://principia-scientific.org/failed-economics-renewable-energy-facts/ **Just read the conclusions.**

and

http://www.timothymaloney.net/Critique_of_100_WWS_Plan.html

and

http://www.sciencedirect.com/science/article/pii/S1364032117304495 - The water, wind and solar, (wws) fantasy

and

Mathijs Beckers' detailed *The Non-Solutions Project.*
https://www.amazon.com/non-solutions-project-non-color/dp/1540804615/ref=asap_bc?ie=UTF8

Chapter 11

The Opposition: Climate Change Deniers, Anti-nuclear Zealots and Profiteers

They are entitled to their own opinions,

but not their own physics.

Donald Trump - "Climate Change is a hoax."

Trump adviser to NASA: No more climate change research.

The Guardian 11-22-16

"Bob Walker, Donald Trump's senior adviser on issues related to the space agency, said... 'NASA's earth science division will have its budget cut, which will reduce its world-renowned research into temperature, ice, clouds and other climate phenomena... NASA should step away from what he previously called 'politically correct environmental monitoring'".

Texas Rep. Louis Gohmert

"God will help us."

Congress is ~ 40% attorneys and 2% scientists/engineers.
There are more Fundamentalists in Congress than scientists.

Helen Caldicott, Barry Commoner, Ralph Nader and others who did good work in ending atmospheric nuclear bomb testing, shifted to being against *everything* nuclear when the testing ended. Unfortunately, their success in limiting CO2-free nuclear power has accelerated Climate Change and aided the expansion environment-damaging "alternatives." Because they have refused to educate themselves on radiation safety, and their incomes are enhanced by promoting radiophobia, they continue to spread distortions and falsehoods. For example, Caldicott always conflates nuclear energy with nuclear bombs even though the two processes are very different.

Well paid anti-nuclear zealots like Hellen Caldicott know that fear is an effective tool for generating support. After the Fukushima accident, she predicted: "...hundreds of thousands of Japanese will be dying within two weeks of acute radiation illness." She also foolishly said that she wouldn't eat food grown in Europe because of radiation from Chernobyl.

Australian author Guy Rundle hysterically predicted, "The Japanese crews will slough their skin and muscles, and bleed out internally under the full glare of the world media".

Caldicott has many prominent critics, both from inside science and without. One of the latter is **George Monbiot**, a respected British journalist and former critic of nuclear power who wrote the following article (edited for length), for the April 5, 2011 issue of the *Guardian*.

The unpalatable truth is that the anti-nuclear lobby has misled us all.

"… The anti-nuclear movement to which I once belonged has misled the world about the impacts of radiation on human health. The claims we have made are ungrounded in science, unsupportable when challenged, and wildly wrong. We have done other people and ourselves a terrible disservice.

"I began to see the extent of the problem after a debate with Helen Caldicott, who is the world's foremost anti-nuclear campaigner. She has received 21 honorary degrees and scores of awards, and was nominated for a Nobel peace prize. Like other greens, I was in awe of her. In the debate, she made some striking statements about the dangers of radiation, so I did what anyone faced with questionable science claims should do: I asked for the sources. Caldicott's response has profoundly shaken me.

"First, she sent me nine documents: articles, press releases and an advertisement. None were scientific articles; none contained sources for the claims she made. But one of the press releases referred to a report by the US National Academy of Sciences, which she urged me to read. I have now done so – all 423 pages. It supports none of the statements I questioned; in fact, it strongly contradicts her claims about the health effects of radiation.

I pressed her further and she gave me a series of answers that made my heart sink – in most cases they referred to publications which had little or no scientific standing, which did not support her claims or which contradicted them. (I have posted our correspondence, and my sources, on my website.)

"For 25 years anti-nuclear campaigners have been racking up the figures for deaths and diseases caused by Chernobyl, and parading deformed babies like a medieval circus. They now claim 985,000 people have been killed by Chernobyl, and that it will continue for many generations to come. These claims are false.

"The U. N. Scientific Committee on the Effects of Atomic Radiation is the equivalent of the IPPC, the Inter-governmental Panel on Climate Change. Like the IPCC, it calls on the world's scientists to read thousands of papers and produce an overview. Here is what it says about the impacts of Chernobyl.

'Of the workers who tried to contain the emergency at Chernobyl, 134 suffered acute radiation syndrome; 28 died soon afterwards. Nineteen others died later, but generally not from diseases associated with radiation. The remaining eighty-seven have suffered other complications, including four cases of solid cancer and two of leukaemia... People living in the countries affected today need not live in fear of serious health consequences from the Chernobyl accident.'

"Caldicott told me that UNSCEAR's work on Chernobyl is "a total cover-up". And though I have pressed her to explain, she has yet to produce even a shred of evidence for this contention….

"Professor Gerry Thomas, who worked on the health effects of Chernobyl for UNSCEAR, tells me there is "absolutely no evidence" for an increase in birth defects. The National Academy paper [that] Dr Caldicott urged me to read came to similar conclusions. It found that radiation-induced mutation in sperm and eggs is such a small risk "'that it has not been detected in humans, even in thoroughly studied irradiated populations such as those of Hiroshima and Nagasaki".

"… Caldicott pointed me to a book which claims that 985,000 people have died as a result of the disaster. Translated from Russian and published by the Annals of the New York Academy of Sciences, this is the only document that appears to support the wild claims made by greens about Chernobyl.

"However, a devastating review in the journal Radiation Protection Dosimetry points out that *the book achieves this figure by assuming that all increased deaths from a wide range of diseases – including many which have no known association with radiation – were caused by the Chernobyl accident.…* The study makes no attempt to correlate exposure to radiation with the incidence of disease.

"Its publication seems to have arisen from a confusion about whether Annals was a publisher or a scientific journal. The academy stated: '*In no sense did*

Annals of the New York Academy of Sciences or the New York Academy of Sciences commission this work; nor by its publication do we intend to independently validate the claims made in translation or in the original publications cited in the work. The translated volume has not been peer reviewed by the New York Academy of Sciences, or by anyone else.'

"Failing to provide sources, refuting data with anecdote, cherry-picking studies, scorning the scientific consensus, invoking a cover-up to explain it: all this is horribly familiar. These are the habits of climate-change deniers...."

Dr. John Kusch, of the Thorium Energy Alliance, has been equally critical:

"Helen Caldicott and Amory Lovins are millionaires who make money from oil companies, coal, natural gas - they are paid to spread fear. Lovins is particularly open and proud of his association with the Petroleum and Gas companies. Their industry is fear and hopelessness... Work by candlelight, don't use toilet paper.... These are pointless and futile. It plays into the money-making, apocalyptic vision they pedal. They know who buys their first-class tickets for their pollution-rich trips to sell their books and give speeches subsidized by the industries they claim to hate.

"...They are business people. Corporate shills of the worst sort who know their clients and customers well, and come through."

According to Rod Adams, Lovins' resume' reveals why his other "accomplishments" don't mean he is an expert on nuclear energy:

"He never completed any disciplined course of study to earn any degree, yet he touted the fact that he was "educated at Harvard and Oxford" for about thirty years. And in about 2006, he started admitting that he had dropped out of both schools.

"His first professional experience in energy issues was working as one of David Brower's campaigners in the UK for the anti-nuclear group Friends of the Earth.

"In 2008, during an interview on Democracy Now, Lovins… admitted that he had worked for oil companies for thirty-five years. That association helps explain his many awards and honors. In 2012, he drew a salary of $725,000 from RMI. (Internal Revenue Service form 990)"

Dr. James Hansen vs Big Green

"I recommend that the public stop providing funds to anti-nuclear environmental groups. Send a letter saying why you are withdrawing your support. Their position is based partly on fear of losing support from anti-nuclear donors, and they are not likely to listen to anything other than financial pressure. If they are allowed to continue to spread misinformation about nuclear power, it is unlikely that we can stop hydro-fracking, continued destructive coal mining, and irreversible climate change."

http://seekerblog.com/2015/03/09/james-hansen-calls-out-big-green-its-the-money-that-drives-their-anti-nuclear-dogma/

http://www.environmentalprogress.org/big-news/2017/3/28/why-the-war-on-nuclear-threatens-us-all

To view a 40-minute excerpt of a video that features real scientists disputing Caldicott and others while exposing their tactics, visit http://youtube/FGYQhGNUMCo. For the full video, see https://www.youtube.com/watch?v=Qaptvhky8IQ

In 2015, anti-nuclear "expert" **Dr. Arjun Makhijani** told a Minnesota Senate Energy Committee that every one of France's nuclear power plants produces "thirty bombs worth of plutonium every year," which is false. (The plutonium produced by France's many reactors is a mixture of isotopes that are even less useful for making bombs than the uranium in the Earth's crust.) Dr. Makhijani also didn't mention the fact that none of the nuclear weapons in world's inventories were produced with plutonium created in civilian nuclear plants.

Organizations like Eco Watch trumpet "... ocean waters off the west coast are testing positive for radioactive elements... Cesium has been detected in seawater having a radio-intensity of **4** Becquerels per cubic meter."

They apparently don't know, or want to admit, that the normal radioactivity of seawater is **12,000** Bq per cubic meter. These people are either fear-mongering or are being willfully ignorant, the latter applying to the prominent head of a Minnesota foundation whose goals I share, but not his passion for windmills and solar arrays.

When I tried to get this "environmentalist" to rethink his enthusiasm for renewables by providing evidence of their faults with polite emails, his response was "stop hassling me."

People like this who talk "planet," but oppose CO_2-free nuclear power, make good livings by promoting carbon-reliant windmill and solar farms, so they have no interest in factual information that challenges their <u>profitable</u> beliefs. They are more devoted to wallets than walruses, and their fingers are in their ears. They simply don't want to know!

The Koch brothers, Coors and most of the carbon companies fund anti-nuclear efforts and employ Climate Change deniers, many of whom worked for companies and organizations like R. J. Reynolds and the Heartland Institute, where they were paid to deliver the corporate line acid rain, tobacco, global warming, overpopulation, and, of course, nuclear power.

However, because solar and wind must be backed up by power plants that largely burn coal or gas, fossil fuel companies sport wind and solar projects, but relentlessly oppose nuclear power because they know it will cripple their profits.

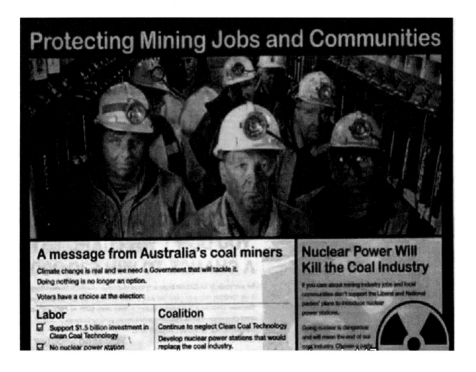

For the coal, oil and gas industries, Profit has always trumped Planet, followed closely by public relations when needed.

For example, BP quickly posted those now-familiar, cheery images of yellow and green blossoms on its gas stations after its Deepwater Horizon disaster soiled the Gulf of Mexico.

Others deceive.

This U.S. government image displays diminishing *tsunami wave heights* following the record-setting earthquake that led to the Fukushima meltdown, but at least one anti-nuclear group claimed that it represented *radiation* spreading across the Pacific Ocean.

http://atomicinsights.com/arnie-gundersen-caught-on-video-lying-about-risk-of-radiation-released-during-fukushima-event/#comment-70328

Nuke Watch, which is written by John La Forge, repeatedly lied about the number of deaths caused by Chernobyl – even after being told that the United Nations Scientific Committee on Effects of Atomic Radiation has counted every death. (43 people had died as of 2004 as a result of radiation exposure at Chernobyl - 28 firefighters "immediately" from radiation; and about 15 more between 1986 and 2004, *perhaps* medically linked to exposure.)

http://www.unscear.org/docs/reports/2008/11-80076_Report_2008_Annex_D.pdf

And when the concentration of Cesium-134 was measured at 0.3 Becquerels per ton of seawater along the coast of Oregon – a miniscule amount - USA Today, the Associated Press, CBS, NBC, and Oregon Public Broadcasting featured that "news" with bold headlines and an ominous-looking Japanese photo of media tourists inspecting Fukushima Daiichi – all of them clad in unnecessary protective gear and face masks. (The natural radiation level of ocean water is approximately 12,000 Bq per ton, and our clueless media were upset about a 0.3 Bq/ton change!)

Still, there is hope. Real environmentalists like Mark Lynas, Patrick Moore, Stephen Tindale, James Lovelock, Stewart Brand and others who previously opposed nuclear power, have become supporters.

Why Nuclear Power Declined

by Carl Wurtz

"Until the late sixties and early seventies, many environmental organizations were pro-nuclear, including the Sierra Club. "Nuclear energy is the only practical alternative that we have to destroying the environment with oil and coal,' said famed nature photographer and Sierra Club Director, Ansel Adams."

"Starting in the mid-sixties, a handful of Sierra Club members feared rising migration into California would destroy the state's scenic character. They decided to attack *all* sources of cheap, reliable power, not just nuclear, in order to slow economic growth.

"If a doubling of the state's population in the next twenty years is to be encouraged by providing the power resources for this growth," wrote David Brower, the Exec. Director of the Sierra Club, "the state's scenic character will be destroyed. More power plants create more industry and greater population density."

"A Sierra Club member named Martin Litton, a pilot and nature photographer for Sunset magazine, led the campaign to oppose Diablo Canyon, a nuclear site where Pacific Gas and Electric proposed to build on the central Californian coast in 1965.

"'Martin Litton hated people,'" wrote a historian about how the environmental movement turned against nuclear. 'He favored a drastic reduction in population to halt encroachment on park land.'" "But the anti-growth activists had a problem: their message was unpopular. So, they shifted their strategy. They worked hard instead to scare the public by preying on their ignorance. Doris Sloan, an anti-nuclear activist, said, 'If you're trying to get people aroused about what is going on... you use the most emotional issue you can find.'

"This included publicizing images of Hiroshima victims and photos of babies born with birth defects. Millions were convinced a nuclear meltdown was the same as a nuclear bomb.

Not Martin Litton. When asked if he worried about nuclear accidents he replied, 'No, I really didn't care. There are too many people anyway.' Why then, all the fear-mongering? 'I think that playing dirty if you have a noble end,' he said, 'is fine.'

"But the fear-mongering worked on a young, renewable energy advocate named Amory Lovins, who began his career crusading against nuclear weapons. Lovins' basic framework of transitioning from nuclear to renewables was promoted by David Brower and Friends of the Earth and was eventually embraced by Sierra Club, Greenpeace, Natural Resources Defense Council, the Union of Concerned Scientists, the German government, Al Gore, and a whole generation of environmentalists.

"The priority of the environmental movement was to phase out nuclear, not fossil fuels. 'It is, above all, the sophisticated use of coal, chiefly at modest scale, that needs development,' Lovins wrote in 1976. Around the same time the Sierra Club's Executive Director, Michael McCloskey, referred to coal as a "bridge fuel" away from nuclear and to renewables."

Dr. Alex Cannara:

"Groups like the Sierra Club, Friends of the Earth, Greenpeace, etc., deserve as much blame as any carbon-seller. They've lied to their members about the safety of nuclear power and avoided educating them about the real environmental hazards that accompany wind and solar."

Even NPR can occasionally slip into "if it bleeds it leads" journalism, which they did when they used the biased title *Fukushima Study Links Children's Cancer to Nuclear Accident* despite the fact that the article contained this statement: "But independent experts say that the study, published in the journal Epidemiology, has numerous shortcomings *and does not prove a link between the accident and cancer.*"

The carbon industry has spent many billions on ads like this ad from the **Oil Heat Institute** that led to the closing of Long Island's $4 billion, Shoreham nuclear power plant, a NEW facility that was ready to generate power.

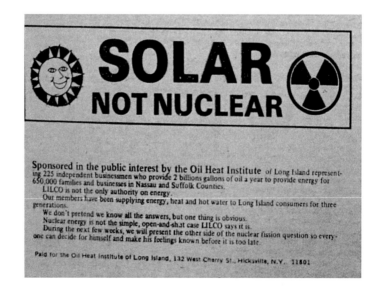

http://atomicinsights.com/smoking-gun-part-18-an-oldie-but-a-goodie-oil-heat-institute-of-long-island-ad-using-scare-tactics-to-fight-shoreham/

Due to Shoreham's closing, thousands of tons of CO_2 and other pollutants have been added to our atmosphere, which has accelerated climate change while the coal, oil and gas industries continue to lie about nuclear power and attempt to paint themselves "green" by promoting carbon-reliant wind and solar power.

Years later, the George W. Bush administration repeatedly tried to censor NASA's James Hansen regarding his presentations and comments about Climate Change, and now the Trump administration has already taken steps to terminate further work on many environmental issues and cripple the exchange of science information.

CENSORING SCIENCE: Inside the Political Attack on Dr. James Hansen and the Truth of Global Warming
MARK BOWEN
Reports of U.S. presidential administrations' successful attempts to discredit and censor scientific evidence about global warming over the past 3 decades have come to light since 2005. James Hansen, director of NASA's Goddard Institute for Space Studies and professor of earth sciences at Columbia University, testified to the Senate in 1988 that global warming required immediate attention. In the years that followed he was repeatedly silenced by government agencies. Hansen's complete story is now told by writer and physicist Bowen, author of *Thin Ice*. The tale covers the science behind climate change, how to mitigate its effect, and the struggle that Hansen faced to get public attention. The book quotes an e-mail sent to Hansen from a NASA administrator that advised: "One way to avoid bad news: stop the measurements!" *Dutton, 2008, 324 p., hardcover, $25.95.*

Chapter 12

Recommended Reading

An Appeal to Reason

"By 2050 we will have added 50% to the world population, which will add 50% more CO2 per year than the eight billion tons we are already adding. Even more alarming was a 2009 release from the National Academy of Science: 'The severity of climate change depends on the magnitude of the change and on the potential for irreversibility. The climate change that takes place due to increases in CO2 is largely **irreversible** for at least 1,000 years after the emissions stop.

"... the prospect of recapturing and sequestering CO_2 from the atmosphere is probably an exercise in futility. Once CO_2 is released, it will take more energy to reclaim it. Unlike our 68,000 tons of nuclear waste, which accounts for just 0.01 % of all industrial toxic waste, there is no place to store the **billions** of tons of CO_2 that will spell disaster within 50 years if we fail to act wisely.

"We must stop using all carbon fuels. Progressively tax energy use. GO NUCLEAR with thousands of on-site MSRs. The power grids we rely on can be damaged, if not destroyed, by a massive solar flare. However, if the U. S. were powered with thousands of LFTRs, these risks would be greatly reduced. Small, modular, inherently safe LFTRs can be built on assembly lines at high speed and shipped by the thousands on semi-trailer trucks."

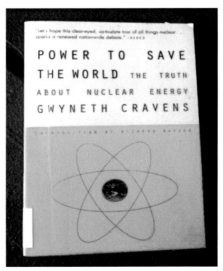

"The power to save the world does not lie in rocks, rivers, wind or sunshine. It lies in each of us."

"Deniers are ideologically committed to attacking an opposing viewpoint – often for financial reasons."

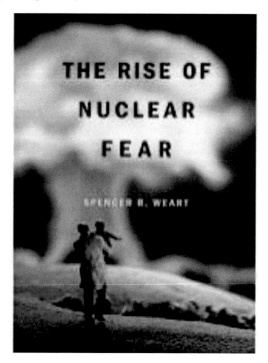

In *The Rise of Nuclear Fear*, Spencer Weart reminds us that many of our "green" organizations, including the Union of Concerned Scientists (UCS), which was formed by East coast, liberals, joined Helen Caldicott in opposing all things nuclear.

"Far from democratic, the biased UCS tolerates little dissent on nuclear matters, which is regrettable because their title gains automatic respect from our unsuspecting public. We have the UCS to blame for the concept of the China Syndrome and the hysteria it promoted when the Three Mile Island accident (in which no one was injured), closely followed the movie of the same name. As a consequence, many nuclear power plant contracts were canceled and replaced by coal-fired plants."

"…Journalists sought out the most worried people to interview, while on national television, Walter Cronkite philosophized about Frankenstein and man's 'tampering with natural forces.' …the *China Syndrome* was just then playing in the theaters. The press, adopting a narrative prepared by the anti-nuclear movement, covered Three Mile Island with an intensity far beyond that accorded to previous industrial accidents. Residents were so upset that some, calling themselves 'survivors', suffered psychological issues…. This was nuclear fear at work, single-minded and unappeasable."

Despite its cartoonish cover, Greenjacked, provides an excellent, expose' of how zealous, science-deficient greens and our fearful public have prevented the expansion of nuclear power, thereby strengthening the climate change they hope to counter with inefficient renewables that are far worse for the environment and less safe than nuclear power.

"The basis for anti-nuclear fear arose from a headline-making theory during efforts to end above ground atom bomb testing. At that time, Nobel prize-winning chemist Dr. Linus Pauling and others used an erroneous "collective dose" theory to prove that fallout would cause thousands of cancers and birth defects.

"Years later, the environmental and peace movements joined forces to block the opening of the Shoreham nuclear power plant, which cost $5.5 billion. Although Jane Fonda and her allies celebrated, few people understood that generating power with coal and gas instead of uranium would cause millions of early deaths and respiratory diseases due to the toxins that burning coal and gas create.

[In contrast, nuclear energy, by displacing the pollution from coal-fired plants, has prevented some 1.8 million premature deaths since 1970." Dr. James Hansen, formerly of NASA.]

"During the seventies, 60 nuclear reactors were planned, but because of the anti-nuclear hysteria caused by Three Mile Island, and later by Chernobyl, that changed when anti-nuclear zealots, who unreasonably tie nuclear weapons to nuclear power, began to dominate environmental organizations. As a result, any attempt to expand nuclear power, our most potent tool for countering climate change, is usually blocked by determined, under-educated people who paint themselves green.

"In 2011, Helen Caldicott began a Montréal press conference by claiming that the Fukushima accident was "orders of magnitude" worse than Chernobyl.

"Orders of magnitude, one of her favorite expressions, means hundreds or thousands of times worse, but it wasn't. It was, however, typical of the rhetoric used by opponents of nuclear power who have little respect for facts.

"And when the twenty-seven United Nations experts who studied the Chernobyl event refuted her claims, Caldicott predictably yelled 'conspiracy and cover-up.'"

Chernobyl provided an excellent example of many environmentalists' disdain for accuracy and the media's willingness to publish unverified claims from dubious sources: One Australian paper quickly trumpeted "2,000 dead," although the death toll through 2016 is closer to fifty.

PLEASE SEE

Popular Science - special ENERGY ISSUE - July, 2011

http://energyrealityproject.com/lets-run-the-numbers-nuclear-energy-vs-wind-and-solar/

http://www.prescriptionfortheplanet.com

http://www.hiroshimasyndrome.com/

http://thoriumenergyalliance.com

An Appeal to Reason

With the harmful effects of Climate Change increasing every year, we must replace carbon-burning power plants with modern, safe, efficient, CO2-free nuclear plants that can consume our stored nuclear waste as fuel, and electrify our vehicles. We must relegate windmills and solar power to remote locations not well served by the grid.

Propelled by physics-ignorant environmentalists and politicians during the last 30 years, we have wasted trillions of dollars on "alternatives" that have added huge amounts of CO2 to our oceans and atmosphere.

There are more than 7 billion humans on earth – far more than our planet can properly support, and that is due in large part to the influence of powerful, anti-birth control religions. These groups will undoubtedly denounce sensible solutions like a proposal from A. J. Shaka: *"Pay people to not have children. Find the price, and pay. It'll be cheaper than any other solution. There is a shot that sterilizes mammals for 10 years. Give it to every 13-year old and then pay people for each additional shot."*

Some say that we have become like cancer cells that are slowly killing their hosts. Cancers, of course, don't know what is coming, but we lack that excuse. It is not too late to adopt effective changes, but we must first overcome our fears and old ways of thinking. Only with nuclear power can we significantly blunt the advance of Climate Change. If we care about our children and the Earth that sustains us, we need to get cracking NOW!

"Terrorism can't and won't destroy our civilization.

Climate Change can and might."

Paul Krugman – N Y Times 11/16/15

We must turn away from carbon.
We must do better than this!

Toles © 2013 The Washington Post.

Reprinted with permission of UNIVERSAL UCLICK.

All rights reserved.

Dr. Erickson is a member of the National Center for Science Education, a past V P of the American Humanist Association, a past President of the Minnesota Humanists and a member of the Thorium Energy Alliance. For more information or to schedule a presentation on nuclear power, radiation safety and alternative energies, please email tundracub@exede.net or caerick@mchsi.com or call 218-744-2003 or 218-744-5182. Please visit www.tundracub.com.

Made in the USA
San Bernardino, CA
01 May 2017